Laboratory Protocols for Mutation Detection

Acknowledgements

The 3rd international workshop on mutation detection, *Mutation Detection 1995,* was held in Visby, Sweden from 18–21 May 1995. The meeting was held under the aegis of HUGO, the Human Genome Organisation, with financial support from Abbott Laboratories, Amersham International Plc, Applied Technology Genetics Corporation, Bio-Rad Laboratories, County Administrative Board of Gotland, European Commission, Federation of European Biochemical Societies, International Science Foundation, Nordic Academy of Advanced Study, Perkin Elmer Applied Biosystems, Pharmacia Biotech AB, Progene Lab AB, Roche Molecular Systems Inc, Sangtec Medical AB, Scandinavian Diagnostic Services, Swedish Cancer Society, Swedish Medical Research Council, Swedish Technical Research Council.

Design of the text layout was the work of Arild Lagerkvist. The cover photograph and the photographs used for section divides were provided by Claes Andersen.

LABORATORY PROTOCOLS FOR MUTATION DETECTION

Edited by

Ulf Landegren
Department of Medical Genetics
Uppsala University
Sweden

Published by Oxford University Press
on behalf of
The Human Genome Organisation

Oxford New York Tokyo
OXFORD UNIVERSITY PRESS
1996

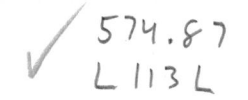

Oxford University Press, Walton Street, Oxford OX2 6DP
Oxford New York
Athens Auckland Bangkok Bombay
Calcutta Cape Town Dar es Salaam Delhi
Florence Hong Kong Istanbul Karachi
Kuala Lumpur Madras Madrid Melbourne
Mexico City Nairobi Paris Singapore
Taipei Tokyo Toronto
and associated companies in
Berlin Ibadan

Oxford is a trade mark of Oxford University Press

Published in the United States by
Oxford University Press Inc., New York

© Oxford University Press, 1996

All rights reserved. No part of this publication may be
reproduced, stored in a retrieval system, or transmitted, in any
form or by any means, without the prior permission in writing of Oxford
University Press. Within the UK, exceptions are allowed in respect of any
fair dealing for the purpose of research or private study, or criticism or
review, as permitted under the Copyright, Designs and Patents Act, 1988, or
in the case of reprographic reproduction in accordance with the terms of
licences issued by the Copyright Licensing Agency. Enquiries concerning
reproduction outside those terms and in other countries should be sent to
the Rights Department, Oxford University Press, at the address above.

This book is sold subject to the condition that it shall not,
by way of trade or otherwise, be lent, re-sold, hired out, or otherwise
circulated without the publisher's prior consent in any form of binding
or cover other than that in which it is published and without a similar
condition including this condition being imposed
on the subsequent purchaser.

A catalogue record for this book is available from the British Library

Library of Congress Cataloging in Publication Data
(Data available)

ISBN 0 19 857795 8

Printed in Great Britain by
Information Press Ltd, Eynsham, Oxon.

Table of contents

CONTRIBUTING AUTHORS

INTRODUCTIONS

   1 Mutation detection now and later
      Ulf Landegren   2

   2 Locus-specific databases
      Richard G.H. Cotton   5

   3 HUGO and the phases of the genome project
      Gert-Jan van Ommen   8

SEARCHING FOR THE PRESENCE OF UNKNOWN MUTATIONS

   4 PCR SSCP - single-strand conformation polymorphism analysis of PCR products
      Kenshi Hayashi   14

   5 SSCP and heteroduplex analysis
      Michael Dean   23

   6 REF & ddF - restriction endonuclease and dideoxy fingerprinting
      Steve S. Sommer   27

   7 DGGE - denaturing gradient gel electrophoresis and related techniques
      Susan E. Murdaugh and Leonard S. Lerman   33

   8 CDCE - constant denaturant capillary electrophoresis for detection and enrichment of sequence variants
      J.S. Hanekamp, P. Andre, H.A. Coller, X.-C. Li, W.G. Thilly, and K. Khrapko   38

   9 LSSP-PCR - multiplex mutation detection using sequence-specific gene signatures
      Sérgio D.J. Pena and Andrew J.G. Simpson   42

SEARCHING FOR THE LOCATION OF UNKNOWN MUTATIONS

10 CCM - chemical cleavage of mismatch
   Susan Ramus and Richard G.H. Cotton   50

11 FAMA - fluorescence-assisted
   mismatch analysis by chemical cleavage
   **Michel Biasotto, Tommaso Meo, Mario Tosi,
   and Elisabeth Verpy**   54

12 Multiplex solid-phase fluorescent chemical cleavage
   **Peter M. Green, Gabriella Rowley, Samia Saad,
   and Francesco Giannelli**   61

13 EMC - enzyme mismatch cleavage
   Rima Youil and Richard G.H. Cotton   65

14 MREC - mismatch repair enzyme cleavage
   A-Lien Lu-Chang   69

DETECTION OF KNOWN MUTATIONS

15 SSO - genetic typing with sequence-specific
   oligonucleotides
   Ulf Gyllensten and Marie Allen   78

16 PASA - PCR amplification of specific alleles
   Steve S. Sommer   82

17 Solid-phase minisequencing
   Ann-Christine Syvänen   87

18 Multiplex solid-phase fluoresecent primer extension
   John M. Shumaker, Andres Metspalu, and C. Thomas Caskey   93

19 OLA - dual-color oligonucleotide ligation assay
   Martina Samiotaki, Marek Kwiatkowski and Ulf Landegren   96

20 LCR - ligase chain reaction
   Bruce Wallace, Luis Ugozzoli, A.A. Reyes, and J. Lowery   101

21 UHG - heteroduplex and
   universal heteroduplex generator analysis
   N.A.P. Wood and J.L. Bidwell   105

DNA SEQUENCE ANALYSIS

22 Solid-phase DNA sequencing
   Mathias Uhlén and Joakim Lundeberg   114

23 Manifold sequencing
   Arild Lagerkvist and Ulf Landegren   119

FLUORESCENT IN SITU HYBRIDIZATION

24 Sensitive FISH using biotin tyramide-based detection
   Ton Raap, Gert-Jan van Ommen, and Joop Wiegant   126

25 Fiber-FISH
   Mervi Heiskanen and Aarno Palotie   131

26 Padlock probes for in situ detection
   Mats Nilsson and Ulf Landegren   135

PROTEIN-LEVEL ASSAYS

17 PTT - protein truncation test
   Rob B. Van Der Luijt, Frans B.L. Hogervorst,
   Johan T. Den Dunnen, P. Meera Khan,
   and Gert-Jan B. van Ommen   140

28 MAMA - monoallelic mutation analysis
   Nickolas Papadopoulos, Ken Kinzler, and Bert Vogelstein 152

NOVEL ASSAY FORMATS

29 Oligonucleotide arrays for scanning nucleic acid sequences
   Edwin Southern, Uwe Maskos, Stephen Case-Green,
   and Martin Johnson   158

30 Optical waveguide device for DNA hybridization analysis
   Julian Gordon, Joanell V. Hoijer, Wang-Ting Hsieh,
   Cynthia Jou, and Don I. Stimpson   164

31 Construction of manifold supports
   Jüri Parik, Arild Lagerkvist, Marek Kwiatkowski,
   and Ulf Landegren   169

MISCELLANEOUS

32 RED - repeat expansion detection
   Martin Schalling, Catherine Erickson-Burgess, Cecilia Zander,
   Kerstin Lindblad, Jeanette Johansen, and Tom Hudson   174

33 Quantitative PCR - competitive PCR followed by
   QPCR detection
   Lucia Cavelier and Ulf Gyllensten   180

34 Capture PCR - amplification with single-sided
   specificity across mutation breakpoints
   Maria Lagerström-Fermér and Ulf Landegren   183

INDEX   190

Contributing authors

MARIE ALLEN (ch.15) Dept of Medical Genetics, BMC, Box 589, S-75123 Uppsala, Sweden. Fax 46 18 526849, E-mail marie.allen@medgen.uu.se

P. ANDRE (ch.8) Center for Environmental Health Sciences, MIT, E18-666, Cambridge, MA 02139, USA. Tel 1 617 253 6220, Fax 1 617 258 5424

MICHEL BIASOTTO (ch.11) Unité d'Immunogénétique & INSERM U 276, Institut Pasteur, 25 rue du Docteur Roux, 75724 Paris Cedex 15, France. Tel 33 1 45 68 85 95, Fax 33 1 40 61 32 36

J.L. BIDWELL (ch.21) Dept of Transplantation Sciences, Molecular Research Div, Homeopathic Hospital Site, Cotham, Bristol BS6 6JU, UK. Tel 44 117 973 0238, Fax 44 117 973 0238, E-mail jeff.bidwell@bris.ac.uk

STEPHEN CASE-GREEN (ch.29) Dept of Biochemistry, University of Oxford, South Parks Road, Oxford OX1 3QU, UK. Tel 44 1865 275 282, Fax 44 1865 275 283

C. THOMAS CASKEY (ch.18) Merck Research Laboratories, Merck & Co, Inc, Summeytown Pike, Room WP26.207, West Point, PA 19486, USA.Tel 1 215 652 7454, Fax 1 215 652 4538

JOHN M. SHUMAKER (ch.18) Dept of Molecular and Human Genetics, Baylor College of Medicine, One Baylor Plaza, Houston, TX 77030, USA

LUCIA CAVELIER (ch.33) Dept of Medical Genetics, BMC, Box 589, S-75123 Uppsala, Sweden. Tel 46 18 174151, Fax 46 18 526849, E-mail lucia.cavelier@medgen.uu.se

H.A. COLLER (ch.8) Center for Environmental Health Sciences, MIT, E18-666, Cambridge, MA 02139, USA. Tel 1 617 253 6220, Fax 1 617 258 5424

RICHARD G.H. COTTON (3,10,13) The Murdoch Institute, Parkville, 3052 Melbourne Victoria Australia. Tel 61 3 345 5045, Fax 61 3 348 139. E-mail cotton@cryptic.rch. unimelb.edu.au

MICHAEL DEAN (ch.5) National Cancer Institute. Frederick Cancer Research and Development Faculty. Frederick MD 21702, USA. Tel 1 301 846 5931, Fax 1 301 846 1909, E-mail dean@fcrfv1.ncifcrf.gov

JOHAN T. DEN DUNNEN (ch.27) MGC-Dept of Human Genetics, Sylvius Laboratory, Leiden University, PO Box 9503, 2300 RA Leiden, The Netherlands. Tel 31 71 276293, Fax 31 71 276075

CATHERINE ERICKSON-BURGESS (ch.32) Dept of Molecular Medicine, Neurogenetics Unit, Karolinska Hospital, S-17176 Stockholm, Sweden. Fax 46 8 327734

FRANCESCO GIANNELLI (ch.12) Paediatric Research Unit, Division of Medical and Molecular Genetics, UMDS, Guy's Hospital, London SE1 9RT, UK, Tel 44 171 955 4450, Fax 44 171 955 4644

JULIAN GORDON (ch.30) Abbott Laboratories, D-90S/AP20, One Abbott Park Road, Abbott Park, IL 60064. E-mail gordonj@apmac.abbott.com

PETER M. GREEN (ch.12) Paediatric Research Unit, Division of Medical and Molecular Genetics, UMDS, Guy's Hospital, London SE1 9RT, UK. Fax 44 171 955 4644, E-mail pgreen@hgmp.mrc.ac.uk

ULF GYLLENSTEN (ch.15,33) Dept of Medical Genetics, BMC, Box 589, S-75123 Uppsala, Sweden. Tel 46 18 174909, Fax 46 18 526849, E-mail ulf.gyllensten@medgen.uu.se

J.S. HANEKAMP (ch.8) Center for Environmental Health Sciences, Massachusetts Institute of Technology, E18-666, Cambridge, MA 02139, USA. Tel 1 617 253 6220, Fax 1 617 258 5424, E-mail hanekamp@mit.edu

KENSHI HAYASHI (ch.4) Division of Genome Analysis, Inst of Genetic Information, Kyushu University, Maidashi 3-1-1, Higashi-ku, Fukuoka 812-82, Japan. Tel 81 92 641 1151, Fax 81 92 632 2375, E-mail khayashi@gen.kyushu-u.ac.jp

MERVI HEISKANEN (ch.25) Dept of Clinical Chemistry, University of Helsinki, Haartmaninkatu 4, 00290 Helsinki, Finland. Tel 358 0 471 4309, Fax 358 0 471 4001, E-mail maheiskanen@cc.helsinki.fi

FRANS B.L. HOGERVORST (ch.27) MGC-Dept of Human Genetics, Sylvius Laboratory, Leiden University, PO Box 9503, 2300 RA Leiden, The Netherlands. Tel 31 71 276293, Fax 31 71 276075

JOANELL V. HOIJER (ch.30) Abbott Laboratories, D-90S/AP20, One Abbott Park Road, Abbott Park, IL 60064, USA

WANG-TING HSIEH (ch.30) Oncor, 209 Perry Parkway, Gaithersburg, MD 20877, USA

TOM HUDSON (ch.32) Whitehead Institute, MIT, Cambridge, Massachusetts, USA

JEANETTE JOHANSEN (ch.32) Dept of Molecular Medicine, Neurogenetics Unit, Karolinska Hospital, S-17176 Stockholm, Sweden. Fax 46 8 327734

MARTIN JOHNSON (ch.29) Dept of Biochemistry, University of Oxford, South Parks Road, Oxford OX1 3QU, UK. Tel 44 1865 275 282, Fax 44 1865 275 283.

CYNTHIA JOU (ch.30) Abbott Laboratories, D-90S/AP20, One Abbott Park Road, Abbott Park, IL 60064, USA

P. MEERA KHAN (ch.27) MGC-Dept of Human Genetics, Sylvius Laboratory, Leiden University, PO Box 9503, 2300 RA Leiden, The Netherlands. Tel 31 71 276293, Fax 31 71 276075

K. KHRAPKO (ch.8) Center for Environmental Health Sciences, MIT, E18-666, Cambridge, MA 02139, USA. Tel 1 617 253 6220, Fax 1 617 258 5424, E-mail khrapko@wccf.mit.edu

KEN KINZLER (ch.28) The Howard Hughes Medical Institute and The Johns Hopkins Oncology Center, 424 North Bond Street, Baltimore, Maryland 21231, USA

MAREK KWIATKOWSKI (ch.19,31) Dept of Medical Genetics, Uppsala Biomedical Center, S-75123 Uppsala, Sweden. Tel 46 18 508760, Fax 46 18 526849, E-mail marek.kwiatkowski@medgen.uu.se

ARILD LAGERKVIST (ch.23,31) Dept of Medical Genetics, Uppsala Biomedical Center, S-75123 Uppsala, Sweden. Tel 46 18 174583, Fax 46 18 526849, E-mail arild.lagerkvist@medgen.uu.se

MARIA LAGERSTRÖM-FERMÉR (ch.34) Dept of Medical Genetics, Uppsala Biomedical Center, S-75123 Uppsala, Sweden. Tel 46 18 174583, Fax 46 18 526849, E-mail maria.lagerström@medgen.uu.se

ULF LANDEGREN (ch.1,19,23,26,31,34) Dept of Medical Genetics, Uppsala Biomedical Center, S-75123 Uppsala, Sweden. Tel 46 18 174910, Fax 46 18 526849. E-mail ulf.landegren@medgen.uu.se

LEONARD S. LERMAN (ch.7) Dept of Biology, MIT, Building 56, Room 731, Cambridge, MA 02139, USA. Tel 1 617 253 6658, Fax 1 617 253 8699, E-mail lerman@fang.mit.edu

X.-C. LI (ch.8) Center for Environmental Health Sciences, MIT, E18-666, Cambridge, MA 02139, USA. Tel 1 617 253 6220, Fax 1 617 258 5424, E-mail xcli@wccf.mit.edu

KERSTIN LINDBLAD (ch.32) Dept of Molecular Medicine, Neurogenetics Unit, Karolinska Hospital, S-17176 Stockholm, Sweden. Fax 46 8 327734

J. LOWERY (p20) Bio-Rad Laboratories, 2000 Alfred Nobel Dr., Hercules, CA 94547, USA. Tel 1 510 741 6532, Fax 1 510 741 1051, E-mail bwallace@bio-rad.com

A-LIEN LU-CHANG (ch.14) Dept of Biological Chemistry, Univ. of Maryland, Baltimore, MD 21201, USA. Fax 1 410 706 1787, E-mail aluchang@umab net.ab.umd.edu

JOAKIM LUNDEBERG (ch.22) Dept of Biochemistry, Royal Institute of Technology, S-100 44 Stockholm, Sweden. E-mail joakiml@biochem.kth.se

UWE MASKOS (ch.29) Dept of Biochemistry, University of Oxford, South Parks Road, Oxford OX1 3QU, UK. Tel 44 1865 275 282, Fax 44 1865 275 283.

TOMMASO MEO (ch.11) Unité d'Immunogénétique & INSERM U 276, Institut Pasteur, 25 rue du Docteur Roux, 75724 Paris Cedex 15, France. Tel 331 45 68 85 98. Fax 331 40 61 32 36, tmeo@pasteur.fr

ANDERS METSPALU (ch.18) Tartu University, Estonian Biocenter, Riia 23, EE2400 Tartu, Estonia. Tel 372 7 420 210, Fax 372 7 420 286, E-mail andres@ebc.ee

SUSAN E. MURDAUGH (ch.7) Dept of Biology, MIT, Building 56, Room 731, Cambridge, MA 02139, USA. Fax 617 253 8699

MATS NILSSON (ch.26) Dept of Medical Genetics, BMC, Box 589, S-75123 Uppsala, Sweden. Tel 46 18 174583, Fax 46 18 526849, E-mail mats.nilsson@medgen.uu.se

AARNO PALOTIE (ch.25) Dept of Clinical Chemistry, University of Helsinki, Haartmaninkatu 4, 00290 Helsinki, Finland. Tel 358 0 471 4309, Fax 358 0 471 4001, E-mail aarno.palotie@hyks.mailnet.fi

NICKOLAS PAPADOPOULOS (ch.28) Howard Hughes Medical Institute and Johns Hopkins Oncology Center, 424 North Bond Street, Baltimore, Maryland 21231, USA

JÜRI PARIK (ch.31) Eesti Biokeskus, Tartu University, Tähetom Toomei, 202400 Tartu. Estonia. Fax 372 2450676, E-mail jparik@ebc.ee

SÉRGIO D. J. PENA (ch.9) 1N˙cleo de Genètica Mèdica de Minas Gerais (GENE/MG), Av. Afonso Pena, 3111/9, 30130-909 Belo Horizonte, MG, Brazil, and Dept. of Biochemistry and Immunology, Federal University of Minas Gerais, C.P. 486, 30161-970 Belo Horizonte, MG, Brazil. Fax 5531 227 3792, E-mail spena@dcc.ufmg.br

TON RAAP (ch.24) MGC-Dept. of Human Genetics, Sylvius Laboratory, Leiden University, P.O. Box 9503, 2300 RA Leiden, The Netherlands. Tel 31 71 276293, Fax 31 71 276075

SUSAN RAMUS (ch.10) The Murdoch Institute, Parkville, 3052 Melbourne Victoria Australia. Fax 613 348 1391, E-mail ramus@cryptic.rch.unimelb.edu.au

A.A. REYES (ch.20) Bio-Rad Laboratories, 2000 Alfred Nobel Dr., Hercules, CA 94547, USA. Tel 1 510 741 6532, Fax 1 510 741 1051, E-mail bwallace@bio-rad.com

GABRIELLA ROWLEY (ch.12) Paediatric Research Unit, Division of Medical and Molecular Genetics, UMDS, Guy's Hospital, London SE1 9RT, UK. Fax 44 171 955 4644

SAMIA SAAD (ch.12) Paediatric Research Unit, Division of Medical and Molecular Genetics, UMDS, Guy's Hospital, London SE1 9RT, UK. Fax 44 171 955 4644

MARTINA SAMIOTAKI (ch.19) Dept of Medical Genetics, Uppsala Biomedical Center, S-75123 Uppsala, Sweden. Tel 46 18 174583, Fax 46 18 526849, E-mail martina.samiotaki@medgen.uu.se

STEVE S. SOMMER (ch.6,16) Dept of Biochemistry and Molecular Biology, Mayo Clinic/Foundation, Rochester, MN 55905 USA. Fax 1 507 284 3383, E-mail sommer.steve@mayo.edu

EDWIN SOUTHERN (ch.29) Dept of Biochemistry, University of Oxford, South Parks Road, Oxford OX1 3QU, UK. Tel 44 1865 275 282, Fax 44 1865 275 283, E-mail ems@biochemistry.oxford.ac.uk

MARTIN SCHALLING (ch.32) Dept of Molecular Medicine, Neurogenetics Unit, Karolinska Hospital, S-17176 Stockholm, Sweden. Fax 46 8 327734, E-mail mschall@gen.ks.se

ANDREW J.G. SIMPSON (ch.9) Ludwig Institute for Cancer Research, R. Prof. Antonio Prudente, 109 - 4th floor, 01509-010 Sao Paulo, SP, Brazil

DON I. STIMPSON (ch.30) Haviland Meteorite Crater, Haviland, KS 67059, USA

ANN-CHRISTINE SYVÄNEN (ch.17) Dept of Human Molecular Genetics, National Public Health Institute, Mannerheimintie 166, SF-00300 Helsinki, Finland. Tel 358 04744270, Fax 358 0 4744480, E-mail christine.syvanen@ktl.fi

W.G. THILLY (ch.8) Center for Environmental Health Sciences, MIT, E18-666, Cambridge, MA 02139, USA. Tel 1 617 253 6220, Fax 1 617 258 5424, E-mail thilly@mit.edu

MARIO TOSI (ch.11) Unité d'Immunogénétique & INSERM U 276, Institut Pasteur, 25 rue du Docteur Roux, 75724 Paris Cedex 15, France. Tel 331 45 68 85 98/5, Fax 331 40 61 32 36, E-mail mtosi@pasteur.fr

LUIS UGOZZOLI (ch.20) Bio-Rad Laboratories, 2000 Alfred Nobel Dr, Hercules, CA 94547, USA. Tel 1 510 741 6532. Fax 1 510 741 1051, E-mail bwallace@bio-rad.com

MATHIAS UHLÉN (ch.22) Dept of Biochemistry, Royal Institute of Technology, S-100 44 Stockholm, Sweden. Tel 1 8 790 7516, Fax 1 8 245 452, E-mail mathias@biochem.kth.se

BRUCE WALLACE (ch.20) Bio-Rad Laboratories, 2000 Alfred Nobel Dr., Hercules, CA 94547, USA. Tel 1 510 741 6532, Fax 1 510 741 1051, E-mail bwallace@bio-rad.com

ROB B. VAN DER LUIJT (ch.27) MGC-Dept of Human Genetics, Sylvius Laboratory, Leiden University, PO Box 9503, 2300 RA Leiden, The Netherlands. Tel 31 71 276293, Fax 31 71 276075

GERT-JAN B. VAN OMMEN (ch.3,24,27) MGC-Dept of Human Genetics, Sylvius Laboratory, Leiden University, PO Box 9503, 2300 RA Leiden, The Netherlands. Tel 31 71 276293, Fax 31 71 276075

ELISABETH VERPY (ch.11) Unité d'Immunogénétique & INSERM U 276, Institut Pasteur, 25 rue du Docteur Roux, 75724 Paris Cedex 15, France. Tel. 331 45 68 85 95, Fax 331 40 61 32 36

BERT VOGELSTEIN (ch.28) Howard Hughes Medical Institute and Johns Hopkins Oncology Center, 424 North Bond Street, Baltimore, Maryland 21231, USA

JOOP WIEGANT (ch.24) MGC-Dept of Human Genetics, Sylvius Laboratory, Leiden University, PO Box 9503, 2300 RA Leiden, The Netherlands Tel 31 71 276293, Fax 31 71 276075

NIGEL A.P. WOOD (ch.21) Dept of Transplantation Sciences, Molecular Research Div Homeopathic Hospital Site, Cotham, Bristol BS6 6JU, UK. Tel 44 117 973 8477, Fax 44 117 973 0238, E-mail nigel.wood@bris.ac.uk

RIMA YOUIL (ch.13) The Murdoch Institute, Parkville, 3052, Melbourne Victoria Australia. Tel 61 3 9 345 5045, Fax 613 348 139

CECILIA ZANDER (ch.32) Dept of Molecular Medicine, Neurogenetics Unit, Karolinska Hospital, S-17176 Stockholm, Sweden, Fax 46 8 327734

INTRODUCTIONS

MUTATION DETECTION
NOW AND LATER
Ulf Landegren

Greatly improved methods for mutation analysis now offer insights into the function of genomes, and allow prediction and diagnosis of disease. Mutation analysis is therefore rapidly assuming central importance in biomedical research as well as in clinical medicine.

A mutation detection workshop

This collection of laboratory protocols for mutation detection techniques arose as a set of recipes shared among participants at a recent HUGO-sponsored workshop, *Mutation Detection '95*, held in Visby, Sweden, May 1995, and organized by Dick Cotton, Ed Southern and myself. An impressive group of originators and major users of methods for mutation detection gathered to exchange experiences and, following the workshop, they were each asked to contribute protocols for the methods that they have developed and are using.

The methods described in this volume span a wide range. Some of the techniques have been widely published and are currently in routine use in many labs while others illustrate principles that may become important in future tests. Some of the methods are easy to establish but may offer less than 100% efficiency, while others are more cumbersome but also more reliable. As described below, the techniques also differ in other respects.

Mutational patterns

While all DNA is subject to mutagenic change, different genes exhibit individual mutational patterns. As a consequence there is a need for many different methods for identifying mutations. For example, in some genes involved in autosomal recessive diseases a limited number of mutations are observed, at least among patients of similar ethnic background. This may be a result of founder effects or it may reflect heterozygote advantage for carriers of the mutations. In other genes, most or all mutations have arisen de novo, although also in this category a predilection for certain sites can sometimes be observed. Long, repetitive genes such as the dystrophin gene may tolerate some missense mutations so a high proportion of large deletions and reading frame-breaking mutations are observed among patients suffering from mutations in genes of this type. In such cases a method that measures the size of the protein product of the gene is effective in detecting the significant mutations. For genes susceptible to mutation by expansion of local trinucleotide repeat motifs special techniques are required both for finding the genes and also for the routine screening of patients. Finally, in both malignancy and infectious diagnosis, it is sometimes important to detect rare sequence variants against a background of the normal sequence.

This protocol collection contains methods which can help to solve the above problems in gene analysis and many others. The information in the following pages should be of value to scientists searching for specific disease genes, and for clinicians examining patients with genetic diseases. The book should also contain some of the nuts and bolts that will be required to build the next generation of genetic tests.

The longitude and latitude of sequence variation

In future medical text books the clinical descriptions of syndromes will undoubtedly be accompanied by lists of all the known genes that may be involved in the diseases. Information on gene variants and their effects on the function of the gene products and the health of the individual is therefore becoming available in the form of mutation databases, as described in a separate foreword by Dick Cotton. Eventually, all genes throughout the human genome will be catalogued in such a manner.

As more and more genes are being investigated, the noise level, represented by polymorphic sequence variants of little or no clinical importance, becomes a consideration. Databases revealing the latitude of normal gene sequence variation may arise as an extension of the Human Diversity Project (which is devoted to studying sequence differences within and between human populations) and this information will be crucial for evaluating the significance of demonstrated sequence variants.

The future of mutation detection

Medicine will increasingly adopt a predictive perspective. In this context, mutation analysis will become a routine aspect of clinical check-ups. Molecular diagnostics offer increased diagnostic precision by enabling disease states to be subdivided according to which genes are involved and the nature of the damage to the genes. Whether a gene product is lost entirely or persists in an altered form may not only influence the character and severity of the disease but may be relevant to the risk of undesired immune reactions during attempts to implement gene therapy.

When expanded triplet repeats were first identified as a cause of genetic disease a few years ago this mutational mechanism came as a surprise. In all likelihood many more mechanisms remain to be disclosed. For instance, as shown recently, changes to patterns of DNA methylation may prove to be an important mechanism in somatic genetic disease. Assays measuring gain- or loss-of-function of gene products can override the requirement to detect specific mutations at the level of DNA, and it is likely that new diagnostic methods will include both gene- and protein-level analyses.

Mutation detection methods evolve rapidly. While the principal means of analysing mutations represented here are likely to remain important, new assay formats and new methods will be developed. Future gene diagnostic strategies can be safely expected to offer vastly increased sample throughput and they will permit rapid screening for mutation in large numbers of genes. Consequently it is increasingly important for each of us to decide what aspects of our genetic fates we wish to know, and how inappropriate access to this information can be avoided.

CHAPTER · 2

LOCUS-SPECIFIC MUTATION DATABASES

Richard G.H. Cotton

We are currently experiencing an explosive increase in the cloning of disease genes as well as in the identification of mutations affecting such genes. This large and sudden expansion of information on mutations has led to many problems with respect to the description and cataloguing of sequence alterations, and to the mechanism for making this information accessible. As a result, up-to-date listings of mutations in genes are just not available for most genes. This has several unfortunate consequences: Scientists have difficulty in establishing whether a particular mutation has been identified previously; biologists do not have access to comprehensive information on functional consequences of different mutations in specific genes; and clinicians are denied the experiences of others who have patients with similar mutations.

Progress towards mutation databases

There are several reasons for the poor accessibility of information on mutations:
1) The rate of discovery of mutations is high and increasing.
2) Many journals refuse to publish single-mutation reports.
3) In contrast to the situation for DNA sequence information, no general acquisition system for mutation reports exist.
4) While a few gene-specific databases have been established, for most genes there are currently no databases.
5) Central databases e.g. OMIM (On-line Mendelian Inheritance in Man) are unable to capture the information because of the sheer volume of mutation reports.

However, some progress has recently been made towards increasing the availability of information on mutations:

1) Several groups of scientists with an interest in the same genes have worked together to create locus-specific mutation databases. Presently, one of the best organised and most well known of these is that for coagulation factor IX, involved in haemophilia B.

2) Some workers have made their mutation databases available over the Internet. Examples of such databases include those for PKU (phenylketonuria), for mutations in factor IX (FIX) and for *P53*, important in malignancy.

3) Central databases such as OMIM have compiled partial listings of mutations that have been identified in specific genes.

Organization of databases

It is clear that the most satisfactory situation would be to have up-to-date annotated listings of mutations in any human genes available via the Internet and an integrated and co-ordinated approach to this objective has recently been initiated. Fundamental to any database system concerning mutations is the nomenclature used for the mutations. Beaudet and Tsui (1993) have started a process towards standardisation of recommendations and it is hoped that recommendations will be agreed upon by the wider "mutation community" and published by the end of 1995.

In October 1994 a meeting was held in Montreal to initiate a move towards uniformity of approach to mutation databases and the possible formation of an Alliance of Database Curators. This approach was approved by Drs. McKusick, Scriver, Kazazian, D. Wallace, Caskey, Tsui and Chakravarti, and has subsequently received the endorsement (in terms of assistance and/or verbal support) of the American Society of Human Genetics (ASHG), the March of Dimes Foundation for Birth Defects, and also the Human Genome Organization (HUGO). The meeting in Montreal resulted in a number of resolutions with respect to mutation databases:

1) Database fields should be defined and kept to a minimum.
2) Models for databases should be generated by C. Scriver and L-C Tsui.
3) Database systems should be simple and driven by curators.
4) Databases should accept both published and unpublished information.
5) Databases should be accessible via GDB on the Internet.

It was also recognized that publishers of relevant journals may have a role in disseminating database information.

These resolutions were further discussed and endorsed at the 3rd international workshop on mutation detection, *"Mutation Detection 1995"*, held in Visby, Sweden, in May 1995 under the auspices of HUGO, which will continue to coordinate developments in the area.

With organizational support from HUGO and funding support from the March of Dimes foundation, a meeting will be held in Minneapolis in October 1995 in association with the annual meeting of ASHG, to further move towards up-to-date on-line mutation databases. The participants at this meeting will include representatives of locus-specific mutation databases, central databases, journals and experts on systems, programs, and the Internet. Various modes of handling information about locus-specific mutations will be discussed, including the following considerations:

1) A central system receiving data via accession number could be established as a requirement for publication, modelled after sequence databases.

2) Locus-specific databases could be managed by curators, down-loading to central databases that are accessible via the Internet, modelled after FIX and p53 at EMBL.

3) The databases could be made available on World Wide Web sites with software available to interrogate all such sites.

We hope that the outcomes of the meeting in Minneapolis will include the formation of an alliance of mutation database curators and the formulation of a plan towards having available on-line up-to-date locus-specific mutation databases.

The three main collaborators in this world-wide endeavour to coordinate the handling of data on locus-specific mutations and ensure that it is readily accessible both to laboratory researchers and clinical practitioners are Dr. V. McKusick, C. Scriver and myself. Anyone interested in any aspect of the project (particularly database curators) is welcome to contact me.

Reference

Beaudet, A.L. and Tsui, L.C. A suggested nomenclature for designating mutations. *Hum. Mutat.* 2: 245-248, 1993.

HUGO
AND THE PHASES OF THE GENOME PROJECT

Gert-Jan van Ommen, HUGO Vice-President for Europe

This book is the result of the 3rd International Workshop on Mutation Detection, held in Visby (Sweden) in May 1995 under the auspices of HUGO. HUGO, the Human Genome Organisation, was established in 1989 as an international organisation for the coordination of human genome research worldwide. It is growing rapidly and currently has nearly 1,000 members from almost 50 countries.

In the first phase of the Human Genome Project, HUGO's activities included overseeing the transition from the biennial series of Human Gene Mapping meetings, the last of which, HGM11, took place in London in 1991, to the current series of Human Genome Mapping (HGM) meetings. The first of the new HGM meetings took place in 1993 in Kobe, Japan. The next, *HGM '96*, will be held in Heidelberg in March 1996 and plans for *HGM '97* in Canada are underway. The large HGM meetings aim to bring together basic and clinical human geneticists and genome scientists, in order to monitor progress in genome mapping and in the understanding of the relationship between gene defects, genome structure and the pathophysiology of genetic disease.

Another important HUGO activity in the genome mapping field has been the organisation and coordination of an extensive world-wide programme of "Single Chromosome Workshops" (SCWs), small meetings held regularly at 1-2 year intervals by the individual chromosome mapping communities to integrate the different large-scale and small-scale mapping efforts ongoing for each chromosome. Thirdly, as a sequel to the long-existing interests of population geneticists in human genetic variation worldwide, HUGO has been involved from the early days in the Human Genome Diversity project, an initiative to describe this variation systematically and install practical and ethical guidelines for this field of research.

Human Genome Workshops

In the second phase of the Human Genome Project, the ongoing identification of our 50-100,000 genes, HUGO's activities have diversified. Alongside the chromosome-by-chromosome approach of the SCW meetings, there is a growing need to address the more and more detailed, trans-genomic issues that are arising. A new program of specialist workshops, "Human Genome Workshops" (HGWs), will be put in place - decisions of the funding agencies allowing - into which the SCW's will gradually be subsumed. Some of the topics for these workshops, such as Comparative Mapping, are not new to human geneticists but they have gathered much more power by the emergence of genome research initiatives in species other than human, notably yeast, *C. elegans*, mouse and lately *Fugu*. Other topics are novel, emerging from our increased insight into genome structure, chromosome make-up and higher order sequence organisation. Thus, increasing attention is being paid to centromeres and telomeres and the interaction of these and other structural elements during gene expression and (dys)regulation. One has only to consider the enigmatic clustering of genes at telomeres, or mechanisms like genomic imprinting or long distance effects on gene expression - for example, in the Prader-Willi/Angelman syndromes and Facioscapulohumeral muscular dystrophy - to realise that genetic pathology does not stop at coding basepairs. The systematic furthering of these fields requires bringing together specialists of diverse backgrounds. Thus, targeted workshops on telomere and centromere structure and long-distance genetic regulation, rank high among HUGO's priorities for the coming years.

Intellectual property issues

The major onslaught on genome sequencing, brought about by the establishment of large-scale, high-throughput automated sequencing laboratories, has precipitated an intense debate on intellectual property issues, associated with the identification of our genetic heritage. The first "HUGO position statement on cDNA patents" was released early in 1992 and later that year HUGO organised an expert workshop on this topic in Munich for the European Commission. Rapid developments and major industrial involvement caused the issue to be revisited at HUGO's "International Genome Summit", held in January 1994 in Houston. This summit brought together for the first time representatives of the governments and funding agencies of most countries sponsoring genome research, together with key specialists in the molecular, technical, clinical, ethical and legal fields. This ultimately resulted in the "HUGO Statement on Patenting of DNA sequences", released in early 1995 in HUGO's regular publication, *Genome Digest*. The statement argued, amongst other points, that sequence data should be immediately and publicly disseminated without restrictions on its use, and that future patent rules, rather than rewarding routine discoveries, should provide protection for the much more intellectually challenging work of determining biological function and application of the gene sequences.

The cDNA/EST initiative

Meanwhile, major industrial and joint industrial/university initiatives are making truly spectacular progress in the identification of expressed sequences, by sequencing cDNA segments from a variety of native and normalized libraries. The Merck-funded programme operated by Washington University, St Louis, has been highly successful in generating publicly accessible data: By November 1995, this single project alone had generated 200,000 of the 270,000 expressed sequence tag (EST) sequences then present in the dbEST database at NCBI in the USA (and mirrored in Europe at the EBI). This treasure trove of raw data greatly stimulates positional cloning and functional analysis worldwide, as witnessed by its attracting more than 12,000 'loggers-on' in one year. To assist in the analysis of this phenomenal data flow, resolve inherent problems of overlap and redundancy, and boost the ultimate goal of mapping these sequences, HUGO has established, as a priority, a short term programme of frequent cDNA/EST mapping workshops, bringing together the active parties and consortia. In 1995 three such workshops took place, in January in London, in May at Cold Spring Harbor and in November in Embiez, near Marseille, with more workshops to follow in 1996. The expectation from the November workshop is that by mid-1996, about 20,000 EST "clusters" (i.e. putative genes) will have been mapped by integrated radiation hybrid and YAC mapping techniques.

Mutation detection

The ultimate goal for defining all our genes is to understand genetic disease - monogenic or multifactorial - at a detailed level, and to assist molecular and clinical geneticists and patients with improved diagnostics, prognostic insights and counselling abilities. This implies that genes not only have to be mapped and sequenced, but that they have to be rendered informative for genetic analysis, in order to follow their segregation in disease families. On one hand this is still being done by linkage analysis, using the vast potential of microsatellite markers and the occasional diallelic polymorphism. On the other hand, however, in recent years we have seen a dramatic improvement in our power to directly detect single basepair mutations and polymorphic sequence variation. This is an area of major clinical impact, driven forward by an active community which, beginning in 1991, has got together every two years. The mutation detection community is growing rapidly in response to the increased opportunities presented by genome technology. For their 1995 meeting, contact was established with HUGO for sponsoring and organisational support. The meeting, the 3rd International Workshop on Mutation Detection, was held in Visby on the island of Gotland, Sweden. Here, nearly 120 senior and junior scientists from 20 countries exchanged the latest tricks of the mutation detection trade in an informal and fruitful atmosphere. An impressive palette of innovation and improvement was presented, from manual methodology to high-tech automation, using biochemical, cytochemical, microtiter, solid state, physicochemical and laser technology. The organisers of the meeting, Richard Cotton, Ulf Landegren and Ed Southern deserve due credit for the excellent and balanced program and for their persistence in hounding the participants for state-of-the-art protocols. The latter were available as a booklet at the meeting, which greatly increased the value of the meeting for the participants. After the meeting, the protocols were deemed to be of immediate value to a much larger community and the decision was made to compile them into a publishable volume. This book is the result, a joint initiative between the meeting organisers, HUGO and Oxford University Press, with special thanks to Ulf Landegren, the editor and driving force behind its appearance. On behalf of HUGO, I express the hope that this book will become a valued asset on many laboratory benches - not only to assist in improved understanding and diagnosis of disease genes, but also to spark off new ideas for even better protocols.

SEARCHING FOR THE PRESENCE OF UNKNOWN MUTATIONS

PCR SSCP
SINGLE-STRAND CONFORMATION POLYMORPHISM ANALYSIS OF PCR PRODUCTS

Kenshi Hayashi

PCR-SSCP is one of the simplest methods for mutation detection. In this method, the target sequence of interest is amplified by PCR and separated as single-stranded molecules by electrophoresis in a non-denaturing polyacrylamide gel (Orita et al., 1989). Sequence variants usually show differences in mobility, and the presence of mutations is revealed as the appearance of new bands in autoradiograms. This mobility shift is believed to be caused by mutation-induced changes of tertiary structure of the single-stranded DNA.

The tertiary structure of single stranded DNA changes under different physical conditions, e.g. temperature and ionic environment. Accordingly, the sensitivity of SSCP depends on these conditions. Although some empirical rules seem to emerge on the choice of separation conditions for sequence variants in particular sequence contexts (Glavac and Dean, 1993), it is not possible to predict whether a certain mutation can be detected under given conditions, especially when the mutation is in a new sequence context. However, the mutation detection sensitivity of PCR-SSCP is generally believed to be high, being more than 80 % in a single run for fragments shorter than 300 bp (Hayashi and Yandell, 1993), although the figure is lower in some publications (Sarkar et al., 1992). Because sensitivity is not 100%, the absence of a new band does not prove that there is no mutation in the analyzed molecule.

Mutations that can be detected by PCR-SSCP are not restricted to fixed positions in the PCR-amplified DNA fragments. Methods in a similar category are DGGE (Fischer and Lerman, 1983), heteroduplex (White et al, 1992) and mismatch cleavage (Winter et al., 1985; Cotton et al., 1988) analyses. Thus, the technique has been used to search for disease-associated unknown mutations in e.g. possibly oncogenic genes in cancer tissues (Suzuki et al., 1990; Gaidano et al., 1991), and in candidate hereditary disease genes in carriers or patients with the particular diseases (Cawthon et al., 1990) to investigate if the genes are indeed responsible. The method is also a valid choice as a diagnostic tool to demonstrate the presence or absence of known mutations, e.g. to determine whether a child carries a mutant allele which has been detected in a parental sample.

The protocol shown below is a modification of the original ones, both of which used ^{32}P as a label for the detection of SSCP bands. The modified protocol includes fewer manipulations and uses less radioactivity. It still has the important advantages of allowing reamplification of each of the separated bands to confirm mutations by direct sequencing of each allele separately (Suzuki et al., 1991). Mutant sequence variants that constitute as little as a few percent of the total number of copies of the sequence can be unambiguously detected and confirmed by this method.

Strategy

The sensitivity of PCR-SSCP decreases with increasing fragment length. Therefore, it is preferable to select primers so that the length of amplification products is less than 300 nucleotides. Mutations in longer DNA molecules (such as whole cDNAs) can be searched by designing primer sets so that the entire region is covered by overlapping short amplification units. Alternatively, long PCR products are digested using the appropriate restriction enzymes and then examined by SSCP. In the latter case it is obviously not possible to reamplify individual bands using the same primer set used for the initial PCR.

As in most PCR-based techniques, specific amplification of target sequences is critical for the unambiguous interpretation of results. Therefore, it is necessary to optimize PCR for the choice of primer sequences and thermal cycling conditions, when new genomic regions are to be examined. For sequences difficult to optimize it may be practical to perform a second, nested PCR to obtain labelled fragments according to the protocol below except with fewer (5 to 10) thermal cycles. In the following protocols, the volume of PCR is minimized for economy and safety. Also, primer and nucleotide concentrations in PCR are lower than generally recommended by suppliers of the enzymes or kits. This is to reduce cost and the hazard of radioactivity, but still ensure high specific activity of label in the PCR product. Low specific activity of the label may lead to overloading of samples in the gel, resulting in reduced resolution. Overloading may also result in complication of the band pattern due to association of primers and single stranded DNA in the sample solution before separation by electrophoresis (Cai and Touitou, 1993).

Once shifted bands are observed in the autoradiogram, these can be excised as small pieces of dried gel, and used for extraction and re-amplification of DNA from single alleles, for subsequent sequencing to identify the mutated nucleotides (Suzuki et al., 1991).

Protocol

PCR using labelled deoxynucleotide triphosphates

Electrophoresis in SSCP-gels

Interpretation of the autoradiogram

Re-amplification of DNA from single SSCP bands and direct sequencing

Variations of PCR-SSCP

PCR using labelled deoxynucleotide triphosphates

In this protocol, PCR products are directly labelled by including (α-^{32}P)-dCTP (10mCi/ml, 3000Ci/mmol, Amersham or NEN) in the amplification reaction.

1) Mix the following solutions (total volume 40.5µl for 10 samples).

H$_2$O	26.0µl
10x PCR buffer	5.0µl
25mM MgCl$_2$	4.0µl
Left primer	1.0µl
Right primer	1.0µl
dNTP	1.0µl
(α-^{32}P)-dCTP	2.0µl
Taq/anti *Taq*.	0.5µl

2) Divide into PCR tubes, 4µl each.

3) Add 1µl of sample DNA (diluted to 50ng/µl with 0.1x TE), and vortex.

4) Overlay with mineral oil (15µl/tube) and briefly spin (if you use a thermal cycler without heated lid).

5) Put the tubes in the thermal cycler and start cycling. Typically, 94°C for the first 1 min, then 30 cycles of 94°C for 30 sec and 60 or 65°C for 2 min.

6) After the PCR, add 45µl of formamide-dye, briefly spin, vortex, and spin again.

Electrophoresis in SSCP-gels

Many physical factors influence the conformation of single-stranded DNA. Among these, the temperature of the gel needs particular attention. Dramatic changes of relative mobilities of sequence variants in SSCP analysis have been demonstrated using perpendicular temperature-gradient gel electrophoresis (Sugano et al., 1995). It is important to use thin gels, such as the ones used in sequencing, in order to avoid temperature rise by ohmic heating. The use of a water-jacketed electrophoretic apparatus (e.g., ATTO Inc.) has the advantages of precise control of gel temperature. Temperatures 10°C below room temperature can be attained by using a cooling plate attached to one side. For running at lower temperature, cooling from both side is required. More economically, two cross-flow (laminar flow) fans, each with a power of about 30W, from both sides of the gel plate is effective in keeping the gel at a constant (room) temperature. With less efficient cooling (such as using conventional turbulent fans), electrophoresis should be run at lower wattage, perhaps 10 to 20W. A sensitive indicator of warming is "smiling" of the band of the leading dye in the gel, bromphenol blue.

Electrophoresis in SSCP is often carried out under two or more conditions to increase detection sensitivity (Hayashi and Yandell, 1993). Gels with 5% glycerol, running at 25°C, are the first choice as this detects most mutations. Additional mutations may be found under other conditions, such as gels without glycerol, running at 25°C.

Use of gels with low cross linker (low ratio of N,N'-methylenebisacrylamide to acrylamide) is important for efficient detection of mutations. The reason for this is unknown. Addition of 5 to 10% glycerol seems to have effects on the mobility similar to, but not exactly the same as, lowering the gel temperature by 10 to 15°C (Hayashi, 1992). The effect of glycerol is primarily to reduce pH by complex formation of glycerol and borate ions in TBE buffer. No glycerol effects are observed in buffers that do not include borate ions. It is possible that lower pH suppresses the charge of phosphate in the nucleic acid backbone, and encourage formation of tertiary structure.

1) Assemble glass plates for the gel.

2) Mix the following solutions in a 50ml Falcon tube. Volumes are for gel of 0.03x30x40cm. Change volumes appropriately for gels of different dimensions.

(50% glycerol	4.5ml, optional)
10x TBE	2.25ml
Acrylamide solution	4.5ml
1.6% ammonium persulfate	1.5ml
H_2O to	45.0ml

3) Add 45μl of TEMED, mix and immediately pour into the gel plate assembly. Keep the plate horizontal, insert the flat side of a shark tooth comb to a depth of approximately 5mm, and leave for at least 2 hr.

4) Remove the comb, put the gel plate in an electrophoresis apparatus (attach aluminum plate if you are using air-cooling system), fill the reservoir with 0.5x TBE, and thoroughly rinse the top surface of the gel with the buffer using a Pasteur pipette. Insert the comb, tooth-side down.

5) Heat the PCR products in formamide dye at 80°C for 5 min, and load the gel (1μl per 5mm lane). Mineral oil (if you are using) need not be removed. Quenching heated samples before loading to gel is not recommended, because it may encourage association of remaining primers and single-stranded DNA and lead to complications in the final SSCP-electrophoretogram (Cai and Touitou, 1993).

6) Start electrophoresis at 40W. Also start cooling by circulating temperature-controlled water to the water-jacket, or by using fans, depending on the instrument. The time required for electrophoresis depends on the length and sequence of the fragment. Suggested times for a first trial is: 1 hr for 150bp fragments when bromphenol blue reaches 5cm from the bottom, and 2 hr for a 400bp fragments when xylene cyanol reaches 5cm from the bottom of the gel without glycerol, running at room temperature. Addition of glycerol slows down the mobility 1.5- to 2-fold. Electrophoresis at lower temperature also requires longer run times.

7) Transfer the gel to a sheet of filter paper, cover with Saran Wrap, and dry using a gel dryer. Contact the dried gel to X-ray film and mark position of contact by stapling at three corners.

8) Expose for a few hr to overnight, depending on radioactivity of the PCR product. Remove staples and develop film.

Interpretation of the autoradiogram

Typically, a PCR-SSCP autoradiogram of one amplification product should give maximally two bands (one each for the separated strands, and a single band if the two strands are not resolved). Sometimes, however, two or more bands appear in SSCP analysis of PCR products representing a single molecular species (cloned DNA, a homozygous locus, etc). This is because the sequence can have more than one stable conformation. The appearance of bands of such iso-conformers is usually reproducible both to position in the gel and relative abundance. Therefore, iso-conformer bands can be easily distinguished from bands of mutant alleles, since control samples analyzed in parallel in the same gel also have these bands.

Re-amplification of DNA from single SSCP bands and direct sequencing.

1) Place filter paper carrying the dried gel (gel-side up) on top of a developed X-ray film, align exactly at the place of contact by referring to the holes of staples, and fix the position again by stapling.

2) Place the film/filter paper/gel on a light box, and wet the appropriate area of the filterpaper with ethanol to make the filter paper half transparent. Cut a small area, e.g. 1x2mm, of gel corresponding to the band of interest using a clean razor blade. Remove Saran Wrap with fine forceps, then peel off the dried gel from the filter paper and drop into 20µl of H_2O in a 0.5ml Eppendorf tube.

3) Heat the tube at 80°C for 3 min, allow to cool to room temperature, and briefly centrifuge to collect all water at the bottom.

4) Take 1µl of the water extract and subject to 25 cycles of PCR with 1µM of each primer (the same primer set used in the first PCR) and 100 to 200µM of each of the four deoxynucleotide triphosphates in 20µl.

5) Add 200µl of H_2O, apply to a Microcon 100 microconcentrator (Amicon) and centrifuge at 500x g (2,500 rpm, r_{center} = 7 cm) for 10 min. Wash twice by adding 200µl of H_2O to the retentate and centrifuging in the same manner.

6) Recover the retentate by fitting the inner cup in an inverted position to a new bottom tube and centrifuge at 3.500rpm (r_{center} = 7cm) for 4 min. Adjust the volume to 20µl by adding H_2O.

7) Estimate DNA concentration by agarose gel electrophoresis. Take 1µl of the cleaned-up PCR product, mix with 3µl of 5% glycerol, 0.05% bromphenol blue and 0.05% xylene cyanol, apply to a 2% agarose mini-gel containing 0.1ppm of ethidium bromide. Also load a known amount of marker DNA, e.g., HaeIII-digested phiX 174 DNA, in a separate lane and separate by electrophoresis. The amount of sample DNA in the gel is estimated by comparing the intensity of the bands of the samples to those of the marker DNA.

8) Use 0.02 to 0.05nmol of the cleaned-up, reamplified DNA for a cycle sequencing reaction. Kits are commercially available (e.g., AmpliCycle, Perkin Elmer; ΔTaq, U.S.B.; SequiTherm, Epicentre).

Variations of PCR-SSCP

Gel matrices other than those described here may enhance separation of mutated strands. Examples are polyacrylamide gels at higher concentrations (Savov et al., 1992) or a new gel matrix, MDE gel (AT Biochem) (Keen et al., 1991). The time required for electrophoresis is longer with these gels, and whether to use them may be a trade-off between efficiency and sensitivity. Agarose is a candidate gel matrix for separation of SSCP-conformers of long DNA fragments. Minisatellite isoalleles (same length but a few base substitutions within the minisatellite repeat) of as long as 6.3 kb could be successfully resolved using agarose gel electrophoresis (Monckton and Jeffreys, 1995).

Several non-radioisotopic PCR-SSCP methods have been reported. These include silver staining of bands (Ainsworth et al., 1991) and use of fluorescent primers in PCR, followed by analysis using an automated DNA sequencer (Makino et al., 1992). Each method has both advantages and disadvantages, as discussed elsewhere (Hayashi and Yandell, 1993).

Reagents

- Primers are stored frozen at -20°C as 10µM solutions in H_2O.
- *Taq*/anti*Taq* is a 1:1 (in volume) mixture of *Taq* DNA polymerase (5U/µl, available from various sources) and Anti-*Taq* antibody™ (1.1mg/ml, *Taq*Start Antibody, Clontech Lab), prepared and stored as recommended by the supplier. The anti-*Taq* antibody suppresses unwanted polymerase action during the preparation of reaction mixture at room temperature (Kellog et al., 1994).
- 10x PK buffer (0.5M Tris-HCl pH 8.3, 0.1M $MgCl_2$ and 50mM dithiothreitol), 10x PCR buffer (0.5M KCl, 0.1M Tris-HCl pH 8.3) and dNTP (a mixture of dATP, dCTP, dGTP and TTP, each at 1.25mM) are stored frozen at -20°C.
- The formamide-dye solution (95% formamide, 20mM Na_2EDTA, 0.05% bromphenolblue and 0.05% xylenecyanol) and the acrylamide solution (49.5% acrylamide and 0.5% N,N'-methylenebisacrylamide) are stored at 4°C.

References

Ainsworth, P.J., Surh, L.C., and Coulter-Mackie, M.B. Diagnostic single strand conformational polymorphism (SSCP): A simple non-radioisotopic method as applied to a Tay-Sachs B1 variant. *Nucl. Acids Res.* **19**: 405-406, 1991.

Botstein, D., White, R.L., and Davis, R.W. Construction of a genetic linkage map in man using restriction fragment length polymorphisms. *Am. J. Hum. Genet.* **32**: 314-331, 1980.

Cai, Q.-Q. and Touitou, I. Excess PCR primers may dramatically affect SSCP efficiency. *Nucl. Acids Res.* **21**: 3909-3910, 1993.

Cawthon, R.M., Weiss, R., Xu, G., Viskochil, D., Culver, M., Stevens, J., Robertson, M., Dunn, D., Gesteland, R., O'Connell, P., and White, R. A major segment of the neuroblastomatosis type 1 gene: cDNA sequence, genomic structure and point mutations. *Cell* **62**: 193-201, 1990.

Cotton, R.G.H., Rodrigues, N.R., and Campbell, R.D. Reactivity of cytosine and thymidine in single-base-pair mismatches with hydroxylamine and osmium tetroxide and its application to the study of mutations. *Proc. Natl. Acad. Sci. USA* **85**: 4397-4401, 1988.

Fischer, S.C. and Lerman, L.S. DNA fragments differing by single base-pair substitutions are separated in denaturing gradient gel. *Proc. Natl. Acad. Sci. USA* **80**: 1579-1583, 1983.

Gaidano, G., Ballerini, P., Gong, J.Z., Inghirami, G., Neri, A., Newcomb, E.W., Magrath, I.T., Knowles, D.M., and Dalla-Favera, R. p53 mutations in human lymphoid malignancies: Association with Burkitt lymphoma and chronic lymphocytic leukemia. *Proc. Natl. Acad. Sci. USA* **88**: 5413-5417, 1991.

Glavac, D. and Dean, M. Optimization of the single-strand conformation polymorphism (SSCP) technique for detection of point mutations. *Hum. Mut.* **2**: 404-414, 1993.

Hayashi, K. PCR-SSCP: A method for detection of mutations. *Genet. Anal. Tech. Applic.* **9**: 73-79, 1992.

Hayashi, K. and Yandell, D.W. How sensitive is PCR-SSCP? *Hum. Mut.* **2**: 338-346, 1993.

Keen, J., Lester, D., Inglehearn, C., Curtis, A., and Bhattacharya, S. Rapid detection of single base mismatches as heteroduplexes on hydrolink gel. *Trends Genet.* **7**: 5, 1991.

Kellog, D. E., Rybalkin, I., Chen, S., Mukhamedova, N., Vlasik, T., Siebert, P.D., and Chenchik, A. *Taq*Start Antibody™: "Hot Start" PCR facilitated by a neutralizing monoclonal antibody directed against *Taq* DNA polymerase. *BioTechniques* **16**: 1134-1137, 1994.

Makino, R., Yazyu H., Kishimoto, Y., Sekiya, T., and Hayashi, K. F-SSCP: A fluorescent polymerase chain reaction-single strand conformation polymorphism (PCR-SSCP) analysis. *PCR Meth. Applic.* **2**: 10-13, 1992.

Monckton, D.G. and Jeffreys, A. Minisatellite isoalleles can be distinguished by single-strand conformational polymorphism analysis in agarose gels. *Nucl. Acids Res.* **22**: 2155-2157, 1995.

Orita, M., Suzuki, Y. Sekiya, T., and Hayashi, K. Rapid and sensitive detection of point mutations and DNA polymorphisms using the polymerase chain reaction. *Genomics* **5**: 874-879, 1989.

Sarkar, G., Yoon, H-S., and Sommer, S.S. Screening for mutations by RNA single-strand conformation polymorphism (rSSCP): comparison with DNA-SSCP. *Nucl. Acids Res.* **20**: 871-878, 1992.

Savov, A., Angelicheva, D., Jordanova, A., Eigel, A., and Kalaydjieva, L. High percentage acrylamide gels improve resolution in SSCP analysis. *Nucl. Acids Res.* **20**: 6741-6742, 1992.

Sugano, K., Fukayama, N., Ohkura, H., Shimosato, Y., Yamada, Y., Inoue, T., Sekiya, T., and Hayashi, K. Single-strand conformation polymorphism analysis by perpendicular temperature-gradient gel electrophoresis. *Electrophoresis* **16**: 8-10, 1995.

Suzuki, Y., Orita, M., Shiraishi, M., Hayashi, K., and Sekiya, T. Detection of ras gene mutations in human lung cancers by single-strand conformation polymorphism analysis of polymerase chain reaction products. *Oncogene* **5**: 1037-1043, 1990.

Suzuki, Y., Sekiya, T., and Hayashi, K. Allele-specific polymerase chain reaction: a method for amplification and sequence determination of a single component among a mixture of sequence variants. *Anal. Biochem.* **192**: 82-84, 1991.

White, M. B., Calvalho, M., Derse, D., O'Brien, S. J., and Dean, M. Detecting single base substitutions as heteroduplex polymorphism. *Genomics* **12**: 301-306, 1992.

Winter, E., Yamamoto, F., Almoguera, C., and Perucho, M. A method to detect and characterize point mutations in transcribed genes: Amplification and overexpression of the mutant c-K-ras allele in human tumor cells. *Proc. Natl. Acad. Sci. USA* **82**: 7575-7579, 1985.

SSCP
AND HETERODUPLEX ANALYSIS
Michael Dean

Single-strand conformation polymorphism (SSCP) analysis is a straightforward technique to distinguish closely similar variants of a given DNA sequence. The technique is used to screen for mutations, since even single nucleotide differences can profoundly influence secondary structure and thus gel migration of the molecule by altering intrastrand basepairing. The ability of the technique to detect mutations, however, depends on a number of factors. Some important factors will be described herein. I also describe a protocol for the heteroduplex technique to scan a DNA segment for mutations.

Protocol

PCR otimization
SSCP reaction
SSCP gel

PCR optimization

Before performing SSCP it is useful to have established the optimal conditions for PCR, using the primers of interest. Important variables to consider are the concentration of Mg^{2+} in the reaction and the parameters of the PCR cycles (temperatures and number of cycles). In addition, several groups have found that a number of additives such as DMSO, spermidine, non-ionic detergents, and tetramethylammoniumchloride may improve the yield of PCR products.

In our hands the Mg^{2+} concentration is the most important variable to be optimized. Set up PCRs with genomic DNA and PCR buffers containing several different concentrations of $MgCl_2$, ranging from 1.5-4.0mM.

PCR in 25μl

1.0μl genomic DNA (100ng)
2.5μl 10x PCR buffer
2.5μl 200μM in each dNTP
1.0μl primer 1 (1OD/ml)
1.0μl primer 2 (1OD/ml)
16.8μl water
0.2μl Taq polymerase (5U/μl)

Make a cocktail of all of the above reagents except the DNA. Add the DNA to the PCR tube and then add 24μl of the cocktail and one drop of mineral oil. Spin briefly in a microfuge and place in a PCR machine.

A standard PCR program is: 95°C 5 min, followed by 30 cycles of 95°C 0.5 min, 55°C 1 min, and 72°C 2 min, with a final extension of 72°C for 10 min.

Following the PCR, 10μl of the reaction can be run on an 8% acrylamide gel or a 2% agarose gel. The products are stained and the conditions giving the highest yield and cleanest products are used for SSCP analysis.

SSCP reaction

The SSCP reaction is identical to the test reaction, except that the concentration of dNTPs is reduced by 1/3 (to 70μM), 0.1μl of $\alpha^{32}P$-dCTP is added to the reaction, and the final volume is 10μl.

10 μl labelled SSCP-PCR
1.0μl genomic DNA (100ng)
1.0μl 10x PCR buffer
0.3μl 200μM each dNTP
1.0μl primer 1 (1OD/ml)
1.0μl primer 2
0.1μl $\alpha^{32}P$-dCTP (3000 Ci/mmol)
5.4μl water
0.2μl *Taq* polymerase (5U/μl)

Other labelled isotopes can be used and both ^{33}P and ^{35}S have been employed. Alternatively the primers can be end-labelled using $\gamma^{32}P$-ATP (note that both primers need to be labelled to visualize both strands.

Two μl of the PCR product is mixed in a well of a microtiter plate with 8μl of loading dye (95% formamide, 0.1% Bromophenol blue, 10mM NaOH). Heat the plate to 95°C for 2 min. Load 2-3μl of the sample onto the gel. It is also useful to load a lane of undenatured DNA, to display the double-stranded DNA fragments.

SSCP gel

SSCP polyacrylamide gels are typically 350mm long and 0.4mm thick and made in 1x TBE. A number of gel formulations and running conditions can be used to vary the separation of different mutations. Important variables are the temperature that the gel is run at, the percent of crosslinking (ratio of acrylamide to bis-acrylamide), and the presence of additives to the gel such as glycerol or sucrose.

- Temperature: Room temperature overnight at 20W constant power, or 50W 3-5 hr in a 4°C cold room.
- Crosslinking: The ratio of acrylamide to bisacrylamide are given in parenthesis: 5%C (19:1), 2.6%C (37.5:1), or 1.3%C (75:1). 0.5x MDE
- Additives: 5-10% glycerol or 10% sucrose

Each of these variables can alter the conformation of single-stranded molecules. Unfortunately, there is no theoretical model which allows prediction of whether a given mutation will be resolved under a specific set of condition. SSCP was initially run on 5% polyacrylamide (5%C) gels at either room temperature or 4°C with or without glycerol. More recently, gels with higher percentage and lower crosslinking have been shown to detect more mutations. Sucrose has also proven to be a useful additive. HydroLink (AT Biochem) gels have also been used for SSCP.

Gel recipes

Final	Stock	ad 250ml
5% acrylamide (1.3%C)	40%	62.5ml
1x TBE	10x	25ml
0.5x MDE	2x	62.5ml
10% glycerol	100%	25ml
1x TBE	10x	25ml
10% acrylamide(1.3%C)	40%	62.5ml
10% sucrose		25g
1x TBE	10x	25ml

40% acrylamide/bis-acrylamide 75:1 (1.3%C)
  39.5g acrylamide
  0.53g bis-acrylamide
  Distilled water to 100ml

- For 75ml of gel solution add 500µl of fresh ammonium persulfate and 50µl of TEMED.
- Dry gels onto Whatman 3MM paper and expose with intensifying screens for 12-72 hr.

Heteroduplex technique for mutation detection

Two complementary DNA strands, derived from alleles that differ in sequence, will include mismatched positions when base-paired. Such double-stranded heteroduplex molecules may show altered migration in native gels, compared to homoduplexes of either allele. This phenomenon forms the basis of a simple technique to screen for mutations.

1) Set up standard PCR incorporating ^{32}P-dCTP during amplification (see above).

2) Dilute the PCR 2-fold with 1x TBE (make mixtures at this point if desired) and heat tubes to 95°C for 5 min and allow to cool to room temperature slowly.

3) Add 8µl of PCR sample to 2µl of nondenaturing loading dye (15% Ficoll, 0.05% xylene cyanol, and 0.05% bromophenol blue) in a microtiter plate.

4) Load 2µl on a 40cm 5% bis-acrylamide/TBE gel and separate by electrophoresis at 2-3W for 12-16 hr at room temperature.

Inspect an autoradiogram of the gel for retarded bands that indicate duplex molecules containing mismatched positions, caused by sequence differences among two coamplified alleles.

References

Dean, M. and Gerrard, B. Helpful hints for the detection of single-stranded conformation polymorphisms. *BioTechniques* **10**: 332-333, 1991.

Glavac, D. and Dean M. Optimization of the single strand-conformation polymorphism (SSCP) technique for detection of point mutations. *Hum. Mut.* **2**: 404-414, 1993.

Ravnik-Glavac, M., Glavac, D., Dean, M. Sensitivity of single-strand conformation polymorphism (SSCP) and heteroduplex method (HA) for mutation detection in the cystic fibrosis gene. *Hum. Mol. Gen.* **3**: 801-807, 1994.

White, M., Carvalho, M., Derse, D., O'Brien, S.J., and Dean, M. Detecting single base substitutions as heteroduplex polymorphisms. *Genomics* **12**: 301-306, 1992.

CHAPTER • 6

REF & ddF
RESTRICTION ENDONUCLEASE AND DIDEOXY FINGERPRINTING

Steve S. Sommer

Single-strand conformation polymorphism (SSCP) is a widely utilized screening method for the detection of mutations due to its simplicity and the general availability of expertise in polyacrylamide gel electrophoresis. However, SSCP does not detect all mutations and its sensitivity can vary dramatically between different regions of DNA. Here I present two alternative modifications of SSCP, both serving to enhance the mutation detection efficiency.

Restriction endonuclease fingerprinting (REF) was developed to detect the presence of essentially all mutations in long, contiguous segments of DNA. Briefly, a DNA segment is amplified by PCR, digested separately with five to six different restriction enzymes, mixed, end-labelled, denatured, and separated by electrophoresis on a nondenaturing gel. If six restriction endonucleases are utilized, a single sequence change is present in six restriction fragments that differ in size and sequence. Upon denaturation, 12 different single-stranded segments are generated. A change in mobility in any one of these segments is sufficient to detect the presence of a mutation. A change in mobility can occur due to a change in secondary structure (SSCP component), or an alteration of restriction sites (restriction component).

Another modification of SSCP, ddF, can detect essentially 100% of mutations by analysis under a single set of conditions. ddF is a hybrid between SSCP and Sanger dideoxy sequencing and it involves nondenaturing gel electrophoresis of Sanger sequencing reaction products, generated using a single dideoxy nucleotide. Blinded analyses indicate that 250 to 300bp segments may be screened with high specificity and essentially 100% sensitivity. Many sequencing protocols can be adapted for ddF. The protocol described herein adapts the genomic amplification with transcript sequencing (GAWTS) protocol for ddF.

REF protocol

PCR amplification
Purification of PCR product
Restriction endonuclease digestions
5'-end labelling of restriction fragments
Electrophoresis
Interpretation of REF autoradiograms

PCR amplification

The PCR mixture contains in a total volume of 150μl: 50mM KCl, 10mM Tris-HCl pH 8.3, 1.5mM $MgCl_2$, 200μM of each deoxyribonucleoside triphosphate (dNTPs), 0.1μM of each primer, 6U of *Taq* DNA polymerase, and 1.2μg of genomic DNA. The cycling parameters are 94°C for 1 min, 55°C for 2 min, 72°C for 3 min, for 30 cycles, followed by a final 10 min elongation at 72°C.

Purification of PCR products

The amplification products are adjusted to a volume of 2ml with TE buffer (10mM Tris-HCl, 0.5mM EDTA pH 8.0) and then purified and concentrated to 5-10μl with a Centricon-100 microconcentrator at 1000x g for 20 min. The amount of recovered DNA is determined by spectrophotometry at 260nm wavelength and then diluted to 20ng/μl.

Restriction endonuclease digestion

The restriction endonucleases are selected from a restriction map of the amplified fragment. One hundred ng of amplified DNA is digested in individual tubes with 5 or 6 restriction endonucleases according to the manufacturer's specifications. Calf intestinal alkaline phosphatase is added to each digest to remove the 5' phosphate of the digested DNA fragments. A final volume of 10μl for each reaction is incubated at 37°C for 6-8 hr.

5'-end labelling of restriction fragments

The restriction enzymes are heat inactivated at 85°C for 20 min. The digestion reactions are combined, heated to 96°C for 5 min, and quick-chilled in an ice water slurry. Five ng of digested DNA products is 5' end-labelled with approximately 6μCi of γ-^{32}P-ATP and 1U of T4 polynucleotide kinase in 50mM Tris-HCl pH 7.4, 10mM $MgCl_2$, 5mM dithiothreitol (DTT), 0.1 mM spermidine, and 1μM unlabelled ATP. Reaction volumes of 2μl and are incubated at 37°C for 45 min.

Electrophoresis

After the labelling reaction, 60μl of stop/loading buffer (7M urea, 50% formamide, 20mM Tris-HCl pH 7.5, 10mM EDTA) is added to each tube. Electrophoresis is performed using a 7.5% GeneAmp and 5.6% polyacrylamide gel in TBE buffer (50mM Tris-borate pH 8.3, 1mM EDTA) at 12 watts constant power. Electrophoresis is performed at either room temperature for 4 hr with the temperature controlled by circulating water at 22-24°C on the glass plate, or at a low temperature achieved by running the gel for 5 hr in a 4°C cold room with a fan directly cooling the plates to 8°C. The gels are dried and subjected to autoradiography.

Interpretation of REF autoradiograms

The autoradiograms are examined for absent, altered, and additional bands. If a novel restriction site is created by a sequence change, two new segments will appear in place of a normal segment. If a restriction site is gained or lost due to a sequence change an altered segment results. This mode of detecting mutations is called the restriction component of REF. Mutations may also affect the migration pattern(s) of one or more of the segments during gel electrophoresis. This is referred to as the SSCP component of REF. The localization of a mutation can be facilitated by including a standard lane on each gel which contains the products of each of the restriction endonuclease digestions.

ddF protocol

PCR amplification

Transcription

End-labelling of primer

Sequencing reaction

Gel electrophoresis

Interpretation

PCR amplification

PCR is performed from human genomic DNA:

1) Add a T7 or SP6 promoter sequence to the 5' end of at least one of the primers to be incorporated into the PCR product:

SP6: 5' ATTTAGGTGACACTATAGAATAG 3'
T7: 5' TAATACGACTCACTATAGGGAGA 3'

2) The PCR mixture contains a total volume of 25μl: 50mM KCl, 10mM Tris-HCl pH 8.3, 1.5-4.5mM MgCl$_2$, 200μM of each dNTP, 0.05-1.0μM of primer, 0.5U of Ampli*Taq*, and 200ng of genomic DNA.

3) Perform 30 cycles of amplification at 94°C for 1 min, 50°C for 2 min, and 72°C for 3 min followed by a 10 min elongation at 72°C.

Transcription

A single stranded RNA template is produced by transcribing the PCR product with T7 or SP6 RNA polymerase:

1) Add a 3μl sample of the amplified material to 17μl of an RNA transcription mixture. The final mixture contains 40mM Tris-HCl pH 7.5, 6mM MgCl$_2$, 2mM spermidine, 10mM NaCl, 0.5mM of the four ribonucleoside triphosphates, 20U RNasin, 10 mM DTT, and 20U of T7 or SP6 RNA polymerase.

2) Incubate samples for 1-2 hr at 37°C and stop the reaction by freezing the sample.

End-labelling of primer

A sequencing primer hybridizing internal to the PCR product is 5' end-labelled according to the following protocol:

1) A 0.1-μg sample of oligonucleotide is incubated in a 13μl reaction volume containing 50mM Tris-HCl pH 7.4, 10mM MgCl$_2$, 5mM spermidine, 100μCi γ-^{32}P-ATP (5000 Ci/mmol), and 10U of T4 polynucleotide kinase for 30 min at 37°C.

2) The reaction is heated to 65°C for 5 min, and 7μl of water is added for a final concentration of 5ng of oligonucleotide per μl.

Sequencing reaction

1) Add a 0.5μl sample of the transcription reaction and 0.5μl of the ^{32}P end-labelled primer to 3μl of the annealing buffer (250mM KCl, 10mM Tris-HCl pH 8.3).

2) Denature the samples at 80°C for 3 min and anneal at 45°C for 15 min.

3) Add 4μl of a mixture containing the following reagents to the primer and RNA template solution: 4μl of reverse transcriptase buffer (24mM Tris-HCl pH 8.3, 16mM MgCl$_2$, 8mM DTT, 0.8mM dATP, 0.4mM dCTP, 0.8mM dGTP, and 1.2mM dTTP) containing 100μg/ml actinomycin D, 1U AMV reverse transcriptase, and 1μl of 0.25-

1.0mM ddNTP (the nucleotide is chosen according to the composition of the template).

4) Incubate this mixture at 55°C for 30 min and stop the reaction by adding 20-100µl of stop buffer (40% formamide, 25mM EDTA, 0.1% bromophenol blue and xylene cyanol FF).

Gel electrophoresis

1) Incubate samples at 94°C for 3 min and place in ice water for 10 min.
2) Load 1µl of sample onto a 48cm long, 0.4mm thick nondenaturing sequencing gel (0.5x MDE) with square sample wells, formed using a square-toothed comb.
3) Separate the samples by electrophoresis at 4°C for 4 hr or room temperature for 3 hr at 13W constant power.
4) After electrophoresis, dry the gel for 1 hr and perform autoradiography.

Interpretation

Mutations are detected by the presence or absence of a band and/or by abnormal migration of one or more bands. The nature of the mutation can be determined by sequencing.

Technical tips and guidelines for ddF

- Choose a dideoxy nucleotide that produces a fairly uniform spacing of termination fragments on a sequencing gel, especially near the top of the gel.
- To avoid false negatives, make sure that the entire region of interest is represented in the ddF gel. Count the number of bands required to cover the region on a sequencing gel, and make sure that at least that many bands are present on the ddF gel.
- Accurate pipetting is critical.
- Use similar concentrations of PCR-amplified DNA for each sample.
- Use square-tooth 64-well combs. Only high-quality wells should be used for loading samples.
- Quick-chill the samples after boiling by immersing in ice water; slow cooling may produce fuzzy bands and erratic cooling may produce artifactual mobility shifts.
- Cool the gel with a fan during electrophoresis or run the gel at a low temperature (4°C).
- Never call a sample negative if it was not run directly next to a sample that has the banding pattern of the control. When a substantial frac-

- tion of samples are expected to have a mutation, process a sufficient number of control samples so that at least one of the two adjacent lanes is very likely to contain a normal sequence.

- Before ascribing a pattern to a mutation, determine that the extension reaction was efficient. Sub-optimal extension reactions give shadow bands that can mimic altered patterns due to mutation. Do not score a sample as positive if the intensity of the signal fades out as the segments get larger; this pattern is due to a poor termination reaction, producing multiple nonspecific termination products by AMV reverse transcriptase.

- To avoid mistaking shadow bands (which are generally faint) for heterozygote mutations, amplify, transcribe, and fingerprint multiple (generally 12 or more) samples at the same time. The intensity of faint shadow bands can vary from reaction to reaction, thereby mimicking a common polymorphism or mutation. A dark exposure of the autoradiographic film will reveal that these bands are present at some level in all the samples.

- If even one band in a SSCP analysis is subtly but clearly different from flanking adjacent normal segments, the sample should be assumed to contain a mutation until high-quality sequence analysis of both strands fails to reveal a mutation.

References

Blaszyk, H., Hartmann, A., Schroeder, J.J., McGovern, R.M., Sommer, S.S., and Kovach, J.S. Rapid and efficient screening for p53 gene mutations by dideoxy fingerprinting. *BioTechniques* **18**: 256-260, 1995.

Liu, Q. and Sommer, S.S. Parameters affecting the sensitivities of dideoxy fingerprinting and SSCP. *PCR Meth. Applic.* **4**: 97-108, 1994.

Liu, Q. and Sommer, S.S. Restriction endonuclease fingerprinting (REF): A sensitive method for screening mutations in long, contiguous segments of DNA. *BioTechniques* **18**: 470-477, 1995.

Sarkar, G., Yoon, H-S., and Sommer, S.S. Dideoxy fingerprinting (ddF): a rapid and efficient screen for the presence of mutations. *Genomics* **13**: 441-443, 1992.

Sommer, S.S., and Vielhaber, E.L. Phage promoter-based methods for sequencing and screening for mutations. In: The polymerase chain reaction, Mullis, K., Ferre, F., and Gibbs, R.A., eds., Birkhäuser, pp. 214-221, 1994.

DGGE
DENATURING GRADIENT GEL ELECTROPHORESIS AND RELATED TECHNIQUES

Susan E. Murdaugh and Leonard S. Lerman*

*Communicating author

Procedures are given here for searching for unknown mutations by denaturing gradient gel electrophoresis in slab gels and for capillary separations under uniform (non-gradient) denaturing conditions (see Hanekamp et al. in this volume). In the former technique, separation of mutant or heteroduplex DNA molecules from wild type is seen under nearly steady-state conditions as differences in the level in the gradient where the bands are located; in the second, separation occurs due to differences in mobility, and running time must be controlled. All procedures based on partial melting (thermal gradients, constant denaturant in slab gels or capillaries, etc.) are similar to one of these, and the details of protocol are easily adapted. It is assumed here that the detailed base sequence is known for applications in which all or nearly all mutants are to be detected. The computer calculation for probe design is an essential part of the protocol.

Protocol

Calculation of probe design

Interpretation

PCR

Gel preparation

Calculation of probe design

For both methods, the fragments of the genomic or transcribed sequence to be examined must be made to conform to effective patterns of thermal stability by selection of endpoints and designing primers to add bases to the ends, if necessary. To determine the region in a fragment in which mutants can be discerned, trial endpoints for PCR primers and the full base sequence are introduced into an easy-to-use computer program, such as MELT87 or its successors (MELT95, to be available from MIT, or MACMELT, based on MELT87, available commercially from BioRad, Hercules, CA).

The resulting plot for a particular fragment, termed a meltmap, shows melting in cooperative regions or domains. Optimal design of a fragment uses end points and end sequences such that:

1) the total fragment length is between about 150 and 500bp,
2) the portion of the sequence under scrutiny lies in the lowest melting domain, which should have a length over 80bp, and
3) the lowest melting domain is bounded at only one end by a region of higher Tm.

More complex or less satisfactory patterns of thermal stability can almost always be converted into favorable fragments by means of one of the following:

1) selection of different fragment endpoints,
2) addition of a 30 or more bp sequence rich in G's and C's, creating a GC-clamp at one end, or by
3) introducing a covalent crosslink between the strands at one end as with a psoralen molecule (Attree et al., 1989).

The most obvious means for adding a clamp is to use a PCR primer carrying a 5' G,C-rich segment (Sheffield et al., 1989). Other means have been described (Abrams et al., 1990, Sheffield et al., 1992). The meltmap is recalculated with the clamp incorporated into the fragment sequence, a simple instruction in MELT95. Conspicuously beneficial effects are usually found from a clamp at one end or the other.

Interpretation

There is a strong correlation between the theoretical meltmap and experimental results. With DGGE the great majority of all base substitutions and sequence alterations within the lowest domain will be seen as a displacement of the fragment band in the gradient. If the test sequence is examined as a heteroduplex with a prototype strand, the gel pattern will show four bands when all of the reassociation products are visible, as with ethidium bromide staining, or two bands when only one strand is labelled. The same GC-clamp primers are used to prepare wild-type fragments for reassociation. If uniform denaturing conditions are used, the selection of optimal denaturant (chemical and/or temperature) is facilitated by the use of the program MUTRAV, which calculates the differences in migration velocity as a function of melting. A difference of 0.1°C is often significant, indicating a narrow optimal range, and calculation in advance obviates lengthy trial and error.

PCR

PCR has been satisfactory using *Pfu* polymerase and the following cycle conditions:

94°C 4 min for 1 cycle; 94°C 30 sec, 55°C 30 sec, 72°C 30 sec for 30 cycles.

Gel preparation

Solutions

1) 20x TAE running buffer: 0.8M Tris base, 0.4M sodium acetate, 20mM Na$_2$EDTA pH to 7.4 with glacial acetic acid. To make 4 liters: 388g Tris base, 218g sodium acetate (131.25g anhydrous), and 30g Na$_2$EDTA. Dissolve in 3.5 liters of distilled water; add glacial acetic acid (120ml) to pH 7.8. Allow to cool to room temperature overnight. Adjust to pH 7.4 and bring to volume.

2) 40% acrylamide/bisacrylamide (37.5/1) stock: 150g of mixed acrylamide in 375ml of distilled water. Deionize with resin beads, filter, and store at 4°C.

3) Gel forming solutions: (100% denaturant is defined arbitrarily as 7M urea and 40% formamide.) Two solutions are needed, high and low denaturant concentrations, differing by at least 10%. These can be made by mixing acrylamide containing 90% denaturant with the same acrylamide concentration and no denaturant. TEMED is added just before making the gradient. 90% stock: formamide 180ml, urea 189g, 40% Acryl:bis 100ml. Bring volume to 475ml with sterile water. Add 10g deionizing resin beads. Stir 10 min. Filter through a 0.45μm nylon filter. Add 25ml 20x TAE buffer. Solutions will last at least three months when stored at 4°C.

4) Polymerization catalysts: Ammonium persulfate (APS), 10% solution in sterile water (100mg/ml). The solution is stable at 4°C for two weeks. N,N,N'N'-Tetramethylethylenediamine (TEMED)- electrophoresis grade, undiluted.

5) Neutral loading solution (6x): 0.025g bromophenol blue, 0.025g xylene cyanol, 4g sucrose. Add 1xTAE buffer to final volume of 10ml. Filter sterilize and store at 4°C.

6) Ethidium bromide: 10 mg/ml stock in water. Use at 0.5-1.0 μg/ml for staining gels.

Pouring gels

The following is taken from previously published papers (Abrams and Stanton, 1992; Fisher and Lerman, 1979) where the gel preparation and equipment has been extensively described. Prospective users should refer to these for full details. Electrophoresis is carried out in a thermostated bath filled with 1x TAE buffer. Efficient vigorous stirring is essential.

I. Parallel gels

We use the SE600 series cassettes from Hoefer Scientific with a 0.75mm spacer. The gel gradients are poured using a simple, two-chamber gradient maker, such as the SG30 from Hoefer, and a variable speed peristaltic pump. For a gradient of 40-65% denaturant, first make dilutions from your stocks for each specified endpoint. We pour 16ml gradients using 8ml of each gelling solution. The remaining space at the top of the gel is filled with 0% denaturant, thereby providing a clear line of demarcation for the gradient.

1) Make appropriate dilutions and store on ice.

2) Add catalysts (1/100 vol 10%APS, 1/200 vol TEMED) and mix by stirring.

3) Pipette the denser solution into the downstream mixing chamber which has a 10x3mm stirbar. Stir on stirplate.

4) Bleed the air out of the connecting chamber.

5) Pipette the less dense solution into the other chamber.

6) Open both the internal and external channels and turn on the peristaltic pump. Pour at a speed of 2.8 ml/min (not more than 10 min).

7) After solutions are exhausted, add 0% denaturant solution (with catalysts added) to the mixing chamber. Layer on top of the gradient to form the wells. The comb should be at a slight angle during pouring, then gently pushed in to a depth of about 10mm, carefully so as not to disturb the gradient.

8) After polymerization, remove the comb and promptly rinse wells with 1xTAE buffer using a syringe.

9) Fill wells with buffer and load samples (1µg total plasmid digest) with a Hamilton syringe.

For standard band separations a parallel gel is used. The bath is routinely maintained at 60.0°C, with 150 volts applied. Calculation by means of the program MUTRAV provides guidance as to the run-time needed for separation of variant fragments according to position of the substitution and the gradient.

II. Perpendicular gels

This approach allows inspection of the full range of denaturing conditions in a comparatively short experiment. Electrophoresis is perpendicular to the gradient, and molecules of the sample migrate through a uniform denaturant concentration. The sample, 2µg of fragment either in 0.25ml 3x loading buffer or in 0.75ml low-melting agarose, is applied to a long well at the top of the gel. The typical run for a 400bp fragment is 4 hr at 150 volts at 60.0°C, following a room temperature run of 40 min at 150V prior to the heated run.

References

Abrams, E.S., Murdaugh, S.E. and Lerman, L.S. Comprehensive detection of single base changes in human genomic DNA using denaturing gradient gel electrophoresis and a GC clamp. *Genomics* 7: 463-475, 1990.

Abrams, E.S. and Stanton, V.P. Jr. Using denaturing gradient gel electrophoresis to study conformational transitions in nucleic acids. In: Methods in Enzymology, Vol. 212 DNA structures, Part B. Chemical and electrophoretic analysis of DNA, edited by Lilley, D.M. and Dahlberg, J.E. San Diego, CA: Academic Press, p. 71-104, 1992.

Attree, O., Vivaud, M., Anselm, S., Lavergne, J.M., and Goossens, M. Mutations in the catalytic domain of human coagulation factor IX: rapid characterization by direct genomic sequencing of DNA fragments displaying an altered melting behavior. *Genomics*, **4**: 266-272, 1989.

Fisher, S.G. and Lerman, L.S. Two-dimensional electrophoretic separation of restriction enzyme fragments of DNA. *Meth. Enzymol.* **68**: 183-191, 1979.

Lerman, L.S. and Silverstein, K. Computational simulation of DNA melting and its application to denaturing gradient gel electrophoresis. *Meth. Enzymol.*, **155**: 482-501, 1987.

Myers, R.M., Maniatis, T. and Lerman, L.S. Detection and localization of single base changes by denaturing gradient gel electrophoresis. *Meth. Enzymol.* **155**: 501-527, 1987.

Sheffield, V.C., Beck, J.S., Stone, E.M. and Myers, R.M. A simple and efficient method for aAttachment of a 40-base pair, G,C-rich sequence to PCR amplified DNA. *Biotechniques* **12**: 386-387, 1992.

Sheffield, V.C., Cox, D.R., Lerman, L.S. and Myers, R.M. Attachment of a 40-base pair G+C rich sequence (GC-clamp) to genomic DNA fragments by the polymerase chain reaction results in improved detection of single-base changes. *Proc. Natl. Acad. Sci. USA* **86**: 232-236, 1989.

This work has been supported by grants from the NIH.

CHAPTER • 8

CDCE
CONSTANT DENATURANT CAPILLARY ELECTROPHORESIS FOR DETECTION AND ENRICHMENT OF SEQUENCE VARIANTS

J.S. Hanekamp*, P. Andre, H.A. Coller, X.-C. LI, W.G. Thilly, and K. Khrapko

*Communicating author

In CDCE, separation of the wild type, mutant, and heteroduplex DNA molecules is achieved by the differential velocity of partly melted DNA species in a medium with uniform denaturant concentration (Khrapko et al., 1994b). DNA fragments suitable for CDCE consist of a 100bp low melting domain flanked by a high melting domain, as in denaturing gradient gel electrophoresis (DGGE). Mutants are most effectively separated from wild type, as in DGGE, as heteroduplexes with a wild type strand.

In CDCE DNA migrates through a 30cm quartz capillary of 75μm inner diameter, filled with a viscous polyacrylamide solution (Khrapko et al., 1994b). A 10cm portion of the capillary, prior to the detector, is heated to a temperature permitting partial melting. Normally, the DNA is fluorescein-labelled and detection is based upon laser induced fluorescence. CDCE allows for the use of smaller sample sizes and provides much higher migration rates than conventional electrophoresis, thereby minimizing the effects of diffusion and increasing resolution. A typical run time is less than 30 min. A sample usually consists of 1×10^8 molecules, and as few as 3×10^4 molecules can be detected in a single peak in the output plot. The matrix in the capillary must be replaced before each run, a procedure which takes about one minute, but the capillary can be re-used hundreds of times. The apparatus is compact and, excluding power supplies, can be mounted on a single small optical bench.

CDCE in combination with high fidelity (hifi) PCR can be used to detect somatic mutations at the frequency found in normal human tissues (Coller and Thilly, 1994; Keohavong and Thilly, 1989; Khrapko et al., 1994a; Thilly, 1985). In this procedure the DNA sample of interest is first restriction digested at sites immediately flanking the primers. The mutants are then converted to heteroduplexes by reassociation with the excess wild type sequences present in the sample. The mutants are enriched relative to wild type on CDCE by collecting, at the end of the capillary, the fraction containing heteroduplexes, followed by hifi PCR of this collected region. The mutant enrichment can be repeated to increase sensitivity, and the mutants are then analyzed by a final CDCE. The present degree of sensitivity of this approach for mutant detection is 1 in 2×10^6 copies of mitochondrial DNA in human tissue samples.

Protocol

Preparation of capillaries

Preparation of polymerized polyacrylamide

Capillary electrophoresis

Preparation of capillaries

A syringe-capillary fitting must be placed on the end of the 100µl high pressure syringe to connect with the capillary. Locally heat one end of a 6cm piece of the tubing on a gas burner, insert the syringe needle 3cm deep, and allow to cool. Then heat the other end and insert the capillary 1cm deep, cool, remove the capillary. A stock supply of capillary is prepared in bulk and stored at room temperature. The inner surface of the capillary must be coated with covalently attached polyacrylamide, which prevents electro-osmotic flow and preserves good resolution.

1) Fill 10 meters of quartz capillary with 100µl of 1M NaOH from a 100µl syringe, incubate 1 hr, wash with 100µl of water.
2) Fill with 100µl of 1M HCl, incubate 10 min, wash with 100µl dry methanol.
3) Fill with 100µl of Bind Silane, incubate overnight.
4) Wash with 100µl methanol.
5) Fill with 100µl of polymerization mixture: 600µl of water, 200µl 30% acrylamide, 200µl 5xTBE, 2.5µl 10% ammonium persulfate, 1µl TEMED.
6) Leave for at least 1 hr before using.

Preparation of polymerized polyacrylamide

The replacement matrix is prepared in large quantities and stored for periods up to several months in 10ml glass syringes at 4°C. From there it can be distributed into 100µl high pressure syringes that are used to replace the matrix in the capillaries. The resolution of the CDCE depends critically upon the properties of the polyacrylamide matrix filling the capillary. Polymerization of acrylamide is carried out after the dissolved oxygen concentration is reduced by argon bubbling, using very low concentrations of initiators. In this way only a few chains are initiated, and those grow to high molecular weight products.

1) Prepare 90ml of 5% acrylamide (BioRad) in standard TBE.
2) Add 30µl of TEMED.
3) Put into a 100ml flask with a stir bar.
4) Put the flask in an ice bath positioned on a stir plate, start stirring.
5) Cover the flask with many layers of Parafilm, insert a long needle and intensely bubble argon for 10 min.
6) Take the needle out of the liquid, but keep it within the flask.
7) Add 30µl of fresh 10% ammonium persulfate, make sure it is well mixed.
8) Immediately start filling 10ml glass syringes with polymerizing solution. The mixture may start getting viscous, thus it is better to use wide bore pipetting needles (Thomas Scientific) to fill the syringes. Be sure to exclude any air bubbles from syringes by pre-rinsing them with small amounts of polymerizing solution.
9) Put the syringes at 0°C in vertical position for several days. Argon bubbles will form in the syringes, but they slowly float up during several days, allowing the lower portion of the matrix to be used. One 10ml syringe will provide enough material for thousands of matrix replacements over several months.

Capillary electrophoresis

DNA fragments are labelled with fluorescein or rhodamine, incorporated at the 5' end of the PCR primer on the high melting domain-side of the fragment.

1) Add the DNA fragments to 4.0µl of a 0.1x TBE buffer solution, in an Eppendorf tube, at a final concentration of 1×10^{10} copies/µl.
2) Load the DNA into the end of the capillary by placing the grounded lead into the solution and applying a current of 2µA for approximately 30 sec, which loads 1×10^8 copies into the capillary. The number of copies loaded into the end of the capillary can be adjusted by changing the loading time.
3) Maintain an electric field of 250V/cm along the capillary. This will cause the fragments to migrate at speeds on the order of 1cm/min.

Comments

- The polyacrylamide matrix within the capillary should be replaced after each run in order to maintain high reproducibility and resolution.
- One determines the best temperature for separation by first starting with the melting temperature of the low melting domain, as determined by a meltmap program, and corrected for denaturant concentration. Test samples of mutants are run at different temperatures close to the melting temperature to optimize resolution. A set of standard fragments with known melting behavior can be labelled with rhodamine and run simultaneously with fluorescein-labelled sample fragments. This requires a two-mode operation of the laser and two independent detectors. The acquired data is stored and analyzed by computer.

Materials

Quartz capillary 75μm inner diameter, 350μm outer diameter (Polymicro Technologies Inc)

High pressure U6K 100μl syringe (Rainin)

Teflon tubing, 1/16" outer diameter, 0.01" inner diameter (Bodman)

Bind Silane (gamma-methacryloxypropyltrimethoxysilane) (Sigma)

References

Coller, H.A. and Thilly, W.G. Development and applications of mutational spectra technology. *Envir. Sci. Technol.* **28**: 478-487, 1994.

Fischer, S.G. and Lerman, L.S. DNA fragments differing by single base-pair substitutions separated in denaturing gradient gels: Correspondence with melting theory, *Proc. Natl. Acad. Sci USA*, **80**: 1579-1583, 1983.

Keohavong, P. and Thilly W.G. Fidelity of DNA polymerases in DNA amplification. *Proc. Natl. Acad. Sci. USA*, **86**: 9253-9257, 1989.

Khrapko, K., Andr, P., Cha, R., Hu, G., and Thilly, W.G. Mutational spectrometry: means and ends. In: Progress in Nucleic Acid Research and Molecular Biology, ed. by K. Moldave **49**: 285-312, 1994a.

Khrapko, K., Hanekamp, J.S., Thilly, W.G., Belenkii, A., Foret, F., and Karger, B.L. Constant denaturant capillary electrophoresis (CDCE): A high resolution approach to mutational analysis. *Nucl. Acids Res.* **22**: 364-369, 1994b.

Thilly, W.G. Potential use of gradient denaturing gel electrophoresis in obtaining mutational spectra from human cells. In: E. Huberman and S.H. Barr, eds., Carcinogenesis, Raven Press, New York, **10**: 511-528, 1985.

LSSP-PCR
MULTIPLEX MUTATION DETECTION USING SEQUENCE-SPECIFIC GENE SIGNATURES

Sérgio D.J. Pena* and Andrew J.G. Simpson

*Communicating author

The low-stringency single specific primer PCR (LSSP-PCR) technique described here was developed to meet the need for a rapid, simple, and precise approach to the identification of DNA fragments, where it is expected that a number of distinct alterations may occur in the sequence of interest and where the precise identification of particular variants of the sequence are sought.

The technique consists in the amplification of a previously purified DNA fragment under very low-stringency conditions, using a single oligonucleotide primer, specific for one of the extremities of the target segment. Productive amplification thus results from the interaction of the primer with its specific complementary site, together with low-specific priming on the other DNA strand at one or more points within the sequence. To ensure that maximal numbers of low-specificity interactions occur, the LSSP-PCR is performed under non stringent hybridization conditions, i.e. low annealing temperature and very high concentrations of both the primer and *Taq* DNA polymerase. Experimental data also provide evidence that during amplification there are interactions between the emerging products which themselves can prime synthesis of fragments, in many cases larger than the original template. The end result of the LSSP-PCR amplification is a highly complex gene signature that is specific for the fragment being probed in which alterations as discrete as single base variants can be detected (Pena et al., 1994).

LSSP-PCR is a technique with high information output in that the end result is a complex electrophoretic pattern where each band is informative (as in DNA sequencing). It thus differs fundamentally from techniques that provide a single piece of information, such as a single altered band or presence or absence of a hybridization signal. We feel that the method is particularly relevant to the identification of genes where the set of possible variants is essentially unlimited, unpredictable, and typically may include multiple polymorphisms in the same molecule. The most obvious examples of such gene sets, in addition to the mitochondrial DNA sequences that we initially studied, are gene fragments from highly variant microorganisms such as viruses e.g. papillomavirus (Villa et al., 1995), bacteria (Leptospira) and protozoa (Trypanosoma cruzi). Thus, LSSP-PCR can act as a powerful means to identify microorganisms, using an initial diagnostic PCR product as a template.

In initial rounds of amplification a limited number of amplified fragments are produced which increase in concentration at an exponential rate (Fig. 1). As these fragments reach critical concentrations they are recruited as primers, resulting in large hybrid amplicons. This cascade takes place at the edge of chaos and is thus, as expected, sensitive to small alterations in the starting conditions. As a result, single base alterations in the initial template can quite radically alter the pattern of the final products.

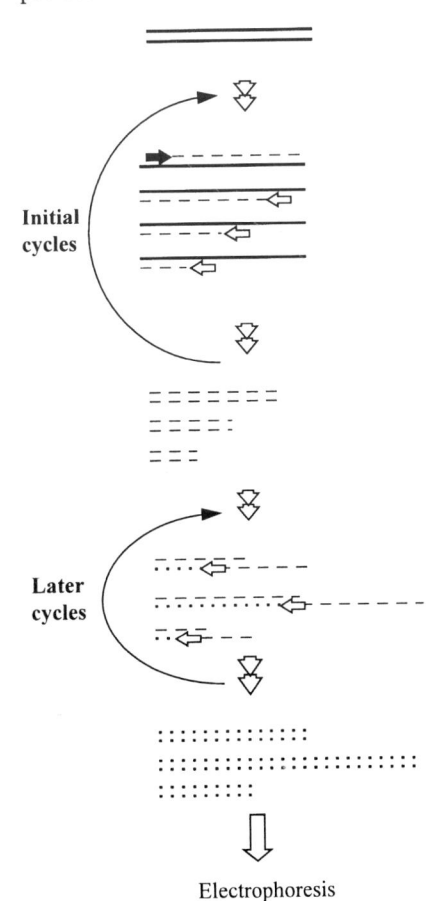

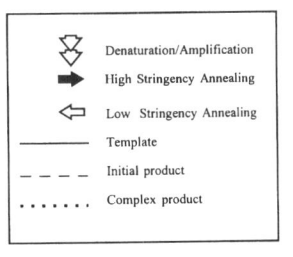

FIGURE 1. General strategy of LSSP-PCR. A purified DNA fragment is submitted to multiple cycles of PCR amplification in the presence of a single oligonucleotide primer, specific for one of the ends of the fragment, under conditions of very low stringency. The primer hybridizes with high specificity to its complementary end and with low specificity to multiple sites within the fragment, in a sequence-dependent manner. In subsequent cycles the initial extension products can themselves act as primers to generate complex hybrid amplicons. The reaction thus yields a large number of products that can be resolved by polyacrylamide gel electrophoresis to give rise to a multiband DNA fragment "signature" that reflects the underlying DNA sequence.

Protocol

Standard LSSP-PCR protocol
Trouble shooting

Standard LSSP-PCR protocol

The equipment and reagents required are those regularly used for PCR.

1) The protocol assumes the availability of a DNA fragment of between 250 and 1,000bp that has probably been produced by PCR. It has been our recent experience that LSSP-PCR results are more reproducible when this first amplification is undertaken under high fidelity conditions, such as those employed for "long PCR" (Barnes, 1994; Cheng et al., 1994), especially when human single copy sequences are being studied.

2) Separate the products in a low-melting agarose gel and stain with ethidium bromide.

3) Excise the band of interest from the gel on a transilluminator (preferably with emission at >300nm to avoid UV-induced nicks), add low TE buffer to give a final concentration of approximately 5-10ng/µl, and heat.

4) Add 1µl of the purified DNA fragment to a PCR containing 40pmol of a single primer (optionally tagged with fluorescein), specific for one of the two ends of the fragment, 1.5U *Taq* polymerase, 200µM dNTPs, 10mM Tris-HCl pH 8.3, 50mM KCl, 1.5mM $MgCl_2$, and 0.001% gelatin, in a reaction volume of 10µl.

5) Cover the reaction with mineral oil and amplify using the following temperature profile: 5 min at 95°C; 40 cycles for one minute each at 95°C, 30°C and 72°C.

6) Separate in a 6% polyacrylamide gel at 100V in TBE buffer until the band with bromophenol blue reaches the end of the gel, or run on an automated fluorescent DNA sequencer.

7) Silver stain according to the following protocol:

   I) 3 min in 10% ethanol, 0.5% acetic acid,
   II) 5 min in 0.2% silver nitrate, 10% ethanol, 0.5% acetic acid,
   III) 2 min wash in Milli-Q water,
   IV) approximately 5 min in 3.0% NaOH, 0.074% formaldehyde until bands appear,
   V) 5 min in 10% ethanol, 0.5% acetic acid,
   VI) transfer to water.

Trouble shooting

- No bands or very simple patterns are produced: probably the enzyme is of reduced activity. Try doubling the amount used, alternatively the primer used is too homologous to an internal sequence; try the other one.
- Poor reproducibility: LSSP-PCR is highly sensitive to sequence alterations and thus artefacts in initial amplification can be detected; use higher concentrations of DNA in the first amplification, repurifying the DNA if necessary.
- Over time all LSSP-PCR patterns for a particular gene grow more similar: the system is contaminated and the original PCR amplification contains a mixture of amplicons: Change all reagents and make sure that amplified DNA products and test DNA samples are kept separate.
- Known polymorphisms are not detected: Check for contamination and try other primers that interact with the template fragment and other fragments containing the region of interest.

Comments

The results obtained using LSSP-PCR are entirely dependent on the template/primer combination being used. They are typically insensitive to small alterations in template concentration. Large template fragments (around 1kb) produce very complex patterns where the majority of the products are smaller than the original template. An example of this kind of result is shown in Fig. 2 where the D-loop of mitochondrial DNA from a mother, child, and father (amplified from peripheral blood using primers L15996 and H408) were examined by LSSP-PCR using primer L15996, and the results analyzed on a polyacrylamide gel. As expected, due to the mitochondrial inheritance of mtDNA, the LSSP-PCR-derived gene signatures from the mother and child are identical, whereas that of the father is distinct. Small templates of around 250bp produce somewhat simpler patterns where typically all the products are larger than the starting template. LSSP-PCR gene signatures produced from smaller templates appear to be less sensitive to point mutations.

LSSP-PCR has three important features: it is extremely simple, it is very sensitive to DNA alterations, and it is potentially highly informative due to its multiband readout. Literally no extra expertise or equipment beyond that required for standard PCR experiments is required to undertake LSSP-PCR analyses, and it can be tried tomorrow in any laboratory with experience of PCR.

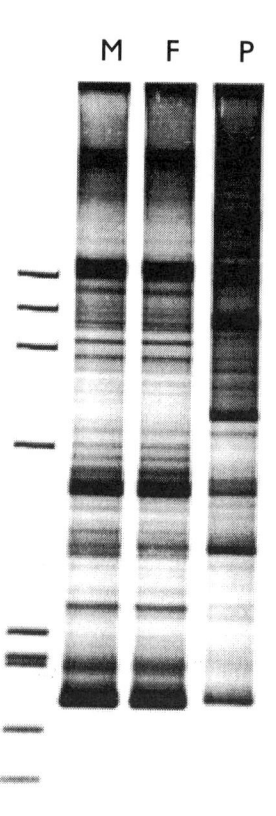

FIGURE 2. A polyacrylamide gel showing fingerprints obtained using the D-loop region of mitochondrial DNA amplified from a trio composed of mother (M), child (F), and father (P). Total genomic DNA was isolated from peripheral blood and a 1,024-bp region of the mtDNA control region of the individuals to be tested was then amplified from 200ng of DNA using the primers L15996 and H408. Ten µl of reaction products were run on a 0.8% low melting point agarose gel, stained with ethidium bromide, and purified by excision from the gel. A sample of the isolated band of approximately 15ng of DNA contained in agarose was then subjected to LSSP-PCR. Following the reaction, 10µl was run on a 6% polyacrylamide gel and silver stained.

The sensitivity of the protocol can also represent a problem in that it apparently detects artefacts introduced by the initial PCR amplification, particularly from complex genomes where the effective concentration of the target is reduced. To overcome the resulting irreproducibility it is important to use as much template as possible which must thus be highly purified and free of inhibitors of the PCR. We thus recommend to mix with the *Taq* polymerase a proof-reading thermostable DNA polymerase for the initial PCR, as is normally done for "long PCR". When small genomes, such as those of microorganisms or mtDNA are used, this problem is much less severe.

The potential information output of LSSP-PCR approaches that of DNA sequencing and in a sense the reaction is a form of sequencing where the banding pattern is sequence-specific but can only be interpreted by reference to known sequence standards. Because of the high level of information it is particularly appropriate for the comparison of DNA fragments where multiple changes occur. This fact coupled with its ready application to small genomes have resulted in the use of the technique in DNA typing, either in the form of human mitochondrial DNA-based identity testing, or for the subclassification of microorganisms.

References

Barnes, W.M. PCR amplification of up to 35-Kb DNA with high fidelity and high yield from bacteriophage templates. *Proc. Natl. Acad. Sci. USA.* **91**: 2216-2220, 1994.

Cheng, S., Chang, S.-Y., Gravitt, P. and Respess, R. Long PCR. *Nature* **369**: 684-685, 1994.

Pena, S.D.J., Barreto, G., Vago, A.R., De Marco, L., Reinach, F.C., Dias Neto, E. and Simpson, A.J.G. Sequence-specific "gene signatures" can be obtained by PCR with single specific primers at low stringency. *Proc. Natl. Acad. Sci. USA* **91**: 1946-1949, 1994.

Villa, L.L., Caballero, O.L., Levi, J.E., Pena, S.D.J. and Simpson, A.J.G. Human papillomavirus identity testing by low stringency single specific primer PCR. *Mol. Cell. Probes* **9**, 45-48, 1995.

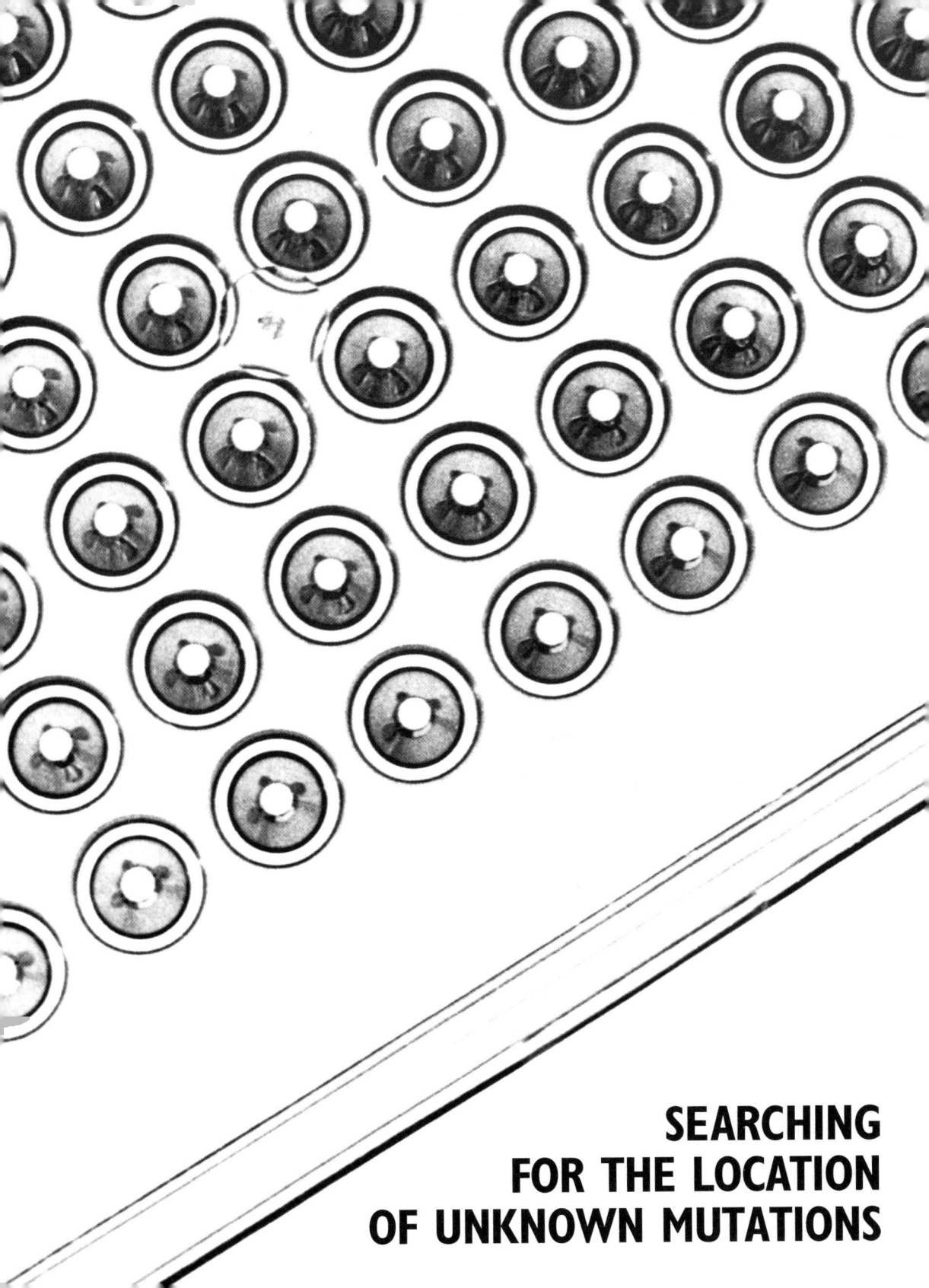

SEARCHING FOR THE LOCATION OF UNKNOWN MUTATIONS

CCM
CHEMICAL CLEAVAGE OF MISMATCH

Susan Ramus and Richard G.H. Cotton*

*Communicating author

The chemical cleavage of mismatch (CCM)-technique for detecting and locating mismatches in heteroduplex DNA molecules (Cotton et al., 1988), relies on the chemical reactivity of mismatched C and T bases to hydroxylamine and osmium tetroxide, respectively. Once reacted, the DNA strands are cleaved at the reacted mismatched base by piperidine and the molecules are separated by size to identify the location of the mismatched positions.

Protocol

Probe preparation
Heteroduplex formation
Chemical modification
Stopping of reactions
Piperidine cleavage

Probe preparation

There are numerous ways to prepare a probe for the CCM method. We find the best results are obtained with probes end-labelled with ^{32}P.

Normal control (wild-type) DNA and patient (mutant) DNA are separately amplified by PCR. The normal PCR product to be end-labelled is eluted from an agarose gel. Approximately 100ng of DNA is end-labelled with γ^{32}P using polynucleotide kinase. The labelled DNA is precipitated twice, and washed with 70% ethanol to remove unincorporated nucleotides. The radioactivity of the probe is determined by counting the dry pellet in a β-counter, and the probe is then resuspended at 1000 cpm/μl in dH$_2$O.

Heteroduplex formation

Labelled heteroduplexes are prepared by adding an excess of unlabelled amplified mutant DNA to the labelled probe. Homoduplexes, used as a control, are prepared by adding unlabelled normal DNA to the probe. For 6 reactions per heteroduplex, take 50μl (50,000cpm) of probe and approximately 500ng of cold DNA, mix and add dH$_2$O to 100μl. One hundred μl of 2x annealing buffer is added, samples are boiled for 5 min, and incubated at 42°C for 1 hr. The heteroduplexes are precipitated and washed, and the dry pellets counted in a β-counter. The heteroduplexes are resuspended in dH$_2$0 at 1000 cpm/μl.

Chemical modification

Six μl (6,000cpm) of the redissolved heteroduplex is aliquoted into separate tubes for either hydroxylamine or osmium tetroxide reactions. Each 6μl of heteroduplex contains approximately 50 to 100 ng of DNA. All further manipulations are carried out in a fume hood.

Hydroxylamine: To 6μl of heteroduplex solution add 20μl hydroxylamine solution (see recipes). Mix well and incubate for 0, 20, or 60 min at 37°C.

Osmium tetroxide: To 6μl of heteroduplex solution add 2.5 μl of 10x osmium tetroxide buffer and 15μl of osmium tetroxide solution (a freshly prepared 1:5 dilution of 4% osmium tetroxide). A yellow colour reaction will occur. Incubate at 37°C for 0, 1, or 5 min.

- All reactions should be done in a fume hood. Collect the wastes separately.

Stopping of reactions

Both the hydroxylamine and osmium tetroxide reactions are stopped by adding 200μl of Hot Stop buffer and then 750μl ice cold 100% ethanol. The excess of single-stranded nucleic acid (tRNA) in the stop buffer effectively stops the chemical modification of mismatches and also helps in the precipitation steps from there on. Chill at -20°C for 30 min and then spin for 15 min. Wash the pellet once with 70% ethanol.

Piperidine cleavage

To each tube add 50μl of 1M piperidine and vortex for 10 sec. Piperidine cleavage is carried out at 90°C for 30 min in a heating block. The osmium tetroxide tubes may turn slightly black at this point. After the cleavage, chill on ice for 2 min, then add 50μl of 0.6M sodium-acetate pH 5.2, followed by 300μl of ice cold 100% ethanol and 2.5μl glycogen (20mg/ml). Chill at -20°C for 30 min and then spin 15 min. Wash pellet once with 70% ethanol, dry and count in a β-counter. The glycogen helps in the precipitation and the resuspension of the pellet in the loading dye.

Samples are resuspended in denaturing dye at 1000cpm/2.5μl. The cleavage products are analysed by electrophoresis of 2.5μl of product on an 8% denaturing polyacrylamide gel with one track containing an end-labelled DNA molecular weight marker (usually *Hae*III-digested phiX174)

Reagents

2x annealing buffer

NaCl (2M)	6ml
Tris-HCl pH 7.5 (1M)	120μl
$MgCl_2$ (1M)	140μl
dH_2O	2.84ml

Hydroxylamine solution

1.39g solid hydroxylammonium chloride (Fluka, Aldrich or BDH Analar) is added to a glass tube and dissolved in 1.6ml water. One ml of diethylamine (Fluka or Aldrich) is added slowly and then another 750μl is added. The pH will be between 6 and 7. Store at 4°C.

- Do not measure the pH directly in the hydroxylamine solution, ie do not put the electrode into the concentrated stock. To determine the pH accurately, remove 2 drops from the stock with a Pasteur pipette and add them to 2ml of water. Measure the pH in this diluted solution.

10x osmium tetroxide buffer

100mM Tris-HCl pH 7.7
10mM EDTA
15% pyridine, use HPLC grade from Aldrich.

4% osmium tetroxide

Osmium tetroxide can be bought from Aldrich already prepared as an aqueous 4% solution. It does, however, lose potency over time (green/greyish discolouration). Loss of the intense yellow colour reaction that occurs upon mixing with pyridine indicates loss of reactivity. Replace it with fresh reagent every 2 months. Store at 4°C.

Hot stop buffer

NaAc pH 5.2 (3M; 1ml)   0.3M

EDTA pH 8.0 (0.1M; 10µl)	0.1mM
tRNA (50mg/ml; 5µl)	25µg/ml
dH$_2$O (9ml)	

Clean up the tRNA (Baker's yeast, Boehringer) before use: phenol and chloroform extract, precipitate with ethanol, and dissolve in water. Heat-treat at 95°C for 10 min and store tRNA stock frozen at 50mg/ml.

1M piperidine

1:10 dilution of piperidine (Fluka) in dH$_2$O (prepare fresh).

Reference

Cotton, R.G., Rodrigues N.R., and Campbell, R.D. Reactivity of cytosine and thymine in single-base-pair mismatches with hydroxylamine and osmium tetroxide and its application to the study of mutations *Proc. Natl. Acad. Sci. USA* 85: 4397-4401, 1988.

FAMA
FLUORESCENCE-ASSISTED MISMATCH ANALYSIS BY CHEMICAL CLEAVAGE

Michel Biasotto, Tommaso Meo, Mario Tosi*, and Elisabeth Verpy

*Communicating author

Mutation scanning methods, based on sizing DNA strands that have been cleaved at mismatched nucleotides, can detect and simultaneously localize mutations in large DNA fragments, longer than in any of the other common methods. The newly introduced fluorescence-assisted mismatch analysis (FAMA) methodology, using bifluorescent DNA substrates, maximizes the reliability and sensitivity of any mutation scanning procedure relying on chemical or enzymatic cleavage of mismatches.

In FAMA complementary strands from wild type and from test DNA are labelled with different fluorophores. As illustrated in Fig. 1, a nested PCR approach can be used to achieve differential fluorescent labelling with high specificity. The target region is PCR-amplified using unlabelled primers. An aliquot thereof is then diluted into a second-round PCR, usually with internally nested fluorescent primers, chosen at a distance of approximately 80bp on either side of the target sequence. Obviously, several other end-labelling methods can be used. Amplicons of a suitable size are chosen, depending on the nature of the project. For example if mismatch scanning is carried out at the mRNA level by RT-PCR, use of a small number of overlapping amplicons, each up to 1.3 kb long, has proven ideal. This target size is close to the upper limit allowed by the electrophoretic conditions presently used for denaturing gels in an ABI 373 DNA sequencer, and the current version of the GENESCAN software.

FAMA permits 100% mutation detection, because each nucleotide substitution yields at least two detectable cleavage products, and because the sensitivity may be greatly improved by highlighting the signal, through superimposition of strand-specific fluorescence chromatograms for different samples. Additional advantages of FAMA are: 1) the convenience of

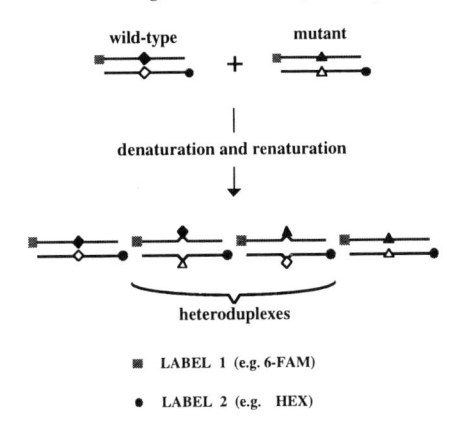

FIGURE 1. Labeling of target amplicons for FAMA. Note that if both wild-type and mutant sequences are present these will be coamplified.

fluorescent labels, which allows different steps involved in the protocol to be separated in time and space; 2) high sensitivity of detection in situations where the mutation is diluted by an excess of wild-type sequence - our data indicate that mutations are efficiently detected at a level of one mutation among ten copies of the sequence; 3) precise localization of the mutation, often at single nucleotide resolution; 4) ability, in many cases, to infer the nucleotide change without sequencing from the specificities of the cleaving reagents and the known sequence of the strand undergoing cleavage; 5) considerable reduction of the time and resources invested in sequencing, by targeting this to a limited region.

Protocol

Generation of fluorescence-labelled template

Heteroduplex formation

Modification reactions

Precipitations

Cleavage of modified bases with piperidine

Loading

Analysis procedure

Generation of fluorescence-labelled template

If the mutation to be identified is expected to be heterozygous, then genomic DNA or reverse transcribed mRNA is amplified by itself. Otherwise an equal amount of reference DNA is included in the PCR. A first PCR is performed on 500ng or less of genomic DNA in a 50µl reaction, using unlabelled oligonucleotide primers with an estimated annealing temperature of 54-56°C according to the Wallace rule, i.e. $(A+T) \times 2°C + (G+C) \times 4°C$. The fluorescence-labelled template is generated by reamplifying 0.5µl of the first amplification reaction in a 100µl reaction, typically for 25 cycles, using internal primers (annealing temperature≈66-70°C) labelled with Fam/Joe or 6Fam/Hex (Perkin Elmer/ABD fluorophores). Try to avoid C or T residues at the three 5'-most positions of these primers and, if necessary, add one or two purines, not present in the target sequence, at the positions close to the fluorophore. Bichrome PCR fragments are ethanol precipitated and quantified on an agarose gel.

Heteroduplex formation

1) For three reactions mix in a screwcap microtube:

DNA	1pmol
10x hybridization buffer	15µl
H_2O	ad 150µl

2) Overlay with two drops of paraffin oil.
3) Boil for 5 min and immediately transfer the tubes to a water-bath at 65°C.
4) Allow to hybridize overnight, or for at least 5 hr.
5) Add 60µg of glycogen and precipitate with 3 volumes of ethanol (20 min in a dry-ice/ethanol bath).
6) Wash once with 70% ethanol.
7) Dry and resuspend the pellets in 18µl H_2O.

Modification reactions

All modification and cleavage reactions are performed in siliconized screwcap O-ring microtubes.

Safety considerations: Osmium tetroxide (OsO_4), diethylamine, pyridine, and piperidine are handled under a fume hood. The supernatant of ethanol precipitation after OsO_4 or hydroxylamine modification are collected for safe disposal.

Hydroxylamine (HA) treatment

Mix heteroduplexes in H_2O, 6µl (0.3pmole), and HA≈5M pH 6.0 (thaw at room temperature), 20µl (final conc≈3.8 M).

1) Incubate at 37°C for 45 min.
2) Transfer to ice, add 200µl cold stop solution.

OsO$_4$ treatment

The final concentrations are: 0.5mM Na$_2$ EDTA, 5mM Hepes pH 8.0, 2 or 2.5% pyridine, 0.4% OsO$_4$ (thaw on ice). Prepare a mix containing H$_2$O, Hepes, Na$_2$ EDTA, and pyridine.

Example of mix for 15 tubes (2.5% pyridine):

Na$_2$ EDTA pH 8.0 (10mM)	18.75µl
Na-Hepes pH 8.0 (100mM)	18.75µl
Pyridine	9.35µl
H$_2$O	88.15µl

Add the mix to all OsO$_4$ reactions on ice. For one tube:

Heteroduplexes	6µl
Mix without OsO$_4$	9µl
1% OsO$_4$	10µl

1) After adding OsO$_4$, mix well by pipetting up and down. The solution turns slightly yellow and small precipitates may appear. These do not affect the activity.
2) Incubate at room temperature for 15 min with gentle agitation in an Eppendorf shaker.
3) Transfer to ice.
4) Add 200µl of cold stop solution.

Precipitations

1) Precipitate the modified DNA with 2.5 vol 100% ethanol at -20°C (565µl). Incubate in a dry-ice/ethanol bath for 20 min. Centrifuge 15 min at 15,000x g or 25 min at 13,000x g (in an Eppendorf centrifuge).
2) Collect supernatants into a HA or OsO$_4$ waste container. Wash the pellets with 1.5ml cold 70% ethanol. Vortex gently, centrifuge at the same speed as above for 5 min and dry. If the supernatant is removed carefully, drying in a Speed Vac is unnecessary.

Comment: If piperidine treatment is not done the same day, precipitate once again, i.e. resuspend the pellets in 200µl of 0.3M sodium acetate, pH 5.2, add 500µl ethanol, and store overnight at -20°C.

Cleavage of modified bases with piperidine

1) Resuspend the pellets in 50µl of 1M piperidine by shaking gently for 45 sec at room temperature in an Eppendorf shaker.

2) Incubate at 90°C for 20 min (screw caps tightly closed).

3) Transfer on ice; open inside the chemical hood.

4) Add 50µl of 0.6M sodium acetate pH 5.2 and precipitate with 250µl ethanol in a dry-ice/ethanol bath for 20 min. Centrifuge as described above.

5) Wash pellets with 1.5ml cold 70% ethanol.

6) Dissolve pellets in 100µl H_2O, freeze the samples in a dry-ice/ethanol bath, and lyophilize. Alternatively: wash once more with 1.5ml cold 70% ethanol, centrifuge 5 min, and dry.

7) Dissolve pellets in 8µl of formamide/EDTA.

8) Store the samples at 4°C until loading. If they are not loaded within 24 hr, store the dry pellets at -20°C.

Gel loading

Mix 4µl (24 lane-gel) or 2µl (36 lane-gel) of the sample in formamide/EDTA with 0.5µl of GS2500 ROX size standards (Applied Biosystems), denature at 94°C for 5 min in a thermocycler or 5 min at 85°C in a waterbath. Keep on ice until loading. Use denaturing gels with acrylamide concentrations between 4% and 6%, depending on the size of the target amplicons, in an Applied Biosystems automated DNA sequencer. Use a square-tooth comb, because, due to the high detection sensitivity, even minor lane-to-lane spill-overs may complicate the analysis. For the same reason, load odd lanes first and run for 5 min. Then load even lanes. We recommend that HA and OsO_4 reactions of the same DNA are run next to each other.

Analysis procedure

Strong cleavage products are seen immediately on the gel window, at the end of the electrophoresis. When scanning for mutations present at a 1:1 ratio with the wild-type sequence, more than 80% of the mutations are easily detected at this stage. Further analysis requires the following steps:

- Distinguishing a weak but true cleavage product from the background.
- Distinguishing between a weak but true cleavage product and spillover from adjacent wells.
- Confirming a mismatch by observing the complementary cleavage product.
- Measuring the size of the cleavage product and deducing, if possible, the exact sequence change.

This analysis is readily performed using the 672 GENESCAN software (Perkin Elmer/ABD). The current version is able to correctly size fragments from slightly below 100bp to 1.3kb. However, in some cases it may be necessary to enter size values as logs of the actual fragment sizes, due to excessive deviation of the mobility of large fragments from proportionality to size. Refer for the analysis to User Bulletin 28 by Perkin-Elmer/ABD: "Fluorescence-Assisted Mismatch Analysis (FAMA) for Mutation Detection" (May 1995).

Materials

7M hydroxylammonium chloride (Merck 4616.0100) is prepared by dissolving 2.43g in H_2O to a final volume of 5ml; 4ml of this solution is titrated to pH 6.0 by addition of diethylamine (Fluka 31.730). The hydroxylamine concentration is then ≈5M. Aliquots can be stored for several months at -80°C in siliconized plastic tubes.

Osmium tetroxide (Aldrich 25,175-5, 4% solution) is diluted in distilled water to give a 1% stock solution, and aliquots are stored at -80°C in siliconized tubes with tight screwcaps.

Pyridine (Aldrich 27,097-0)

Piperidine (Aldrich 10,409-4). Prepare just before use a 1M solution in distilled H_2O. Alternatively, aliquots of 1M piperidine can be stored at -80°C

10x hybridization buffer: 3M NaCl, 35mM $MgCl_2$, 30mM Tris-HCl pH 7.7

Glycogen (Boehringer 901393) 20 mg/ml

Na_2 EDTA pH 8.0, 10mM

Na-Hepes pH 8.0, 100mM

Stop solution: 0.1mM Na_2 EDTA, 0.3M sodium acetate pH 5.2. tRNA may be added to a final concentration of 50µg/ml to stop the modification reactions more effectively. Note however that in this protocol glycogen is used for precipitation.

Sodium acetate pH 5.2 0.6M

Formamide/EDTA: Mix 5 volumes of deionized formamide with one volume of 50mM Na_2 EDTA, pH 8.0, store at -20°C.

Use Eppendorf tubes with screw caps (e.g. Sarstedt, P/N 72.692.100).

GS 2500 ROX size standards (Perkin Elmer/ABD P/N 401100)

Reference

Verpy, E., Biasotto, M., Meo, T. and Tosi, M. Efficient detection of point mutations on color-coded strands of target DNA. *Proc. Natl. Acad. Sci. USA* **91**: 1873-1877, 1994.

CHAPTER · 12

MULTIPLEX SOLID PHASE FLUORESCENT
CHEMICAL CLEAVAGE

Peter M. Green*, Gabriella Rowley, Samia Saad,
and Francesco Gianelli

*Communicating author

The chemical cleavage of mismatch-method may routinely screen PCR segments of at least 1.6kb, it approaches 100% detection and, importantly, it precisely locates mutations within the segments analysed (Cotton et al., 1988; Montandon et al., 1989; Naylor et al., 1991). These properties have made this technique extremely useful for the analysis of large and complex genes in combination with RT-PCR from transcripts of the genes. If the tissue where the gene in question is known to express is not readily available, so-called illegitimate transcripts may be studied in more accessible tissues such as peripheral blood leukocytes (Naylor et al., 1991; Roberts et al., 1992). We have increased by at least 10-fold the speed of chemical mismatch detection by introducing fluorescent labelling, which allows multiplexing (Haris et al., 1994), and solid phase chemistry (Rowley et al., in press).

Introduction of fluorescent tags in the primers used to amplify the probe DNA, and the four-colour fluorescence scanning provided by the ABI 373 DNA Sequencer, allow the concurrent analysis of four DNA segments in a single mismatch reaction and gel lane. Furthermore, the alternative option of internal fluorescent label incorporation increases the flexibility of the method. Solid phase chemistry eliminates the most tedious manipulations of the method and allows the use of a microtitre plate format to screen up to 192 (48 x 4 colours) segments in a single plate and using 3 polyacrylamide gels. The two alternative methods of labelling are presented below as a) end-labelling, and b) internal labelling.

Protocol

Preparation of probe DNA
Preparation of target DNA
Chemical cleavage of mismatch
Analysis of gels

Preparation of probe DNA

End-labelled probe DNA

Sequences from a normal individual (or a known mutant as a positive control) are amplified under standard conditions but using two fluorescence-labelled primers. For economy, these primers can be used at about 1ng/µl final concentration. Products are then gel purified and resuspended in 1/5th the original volume of TE, and can be stored at 4°C.

Internally-labelled probe DNA

Internally labelled probes are prepared using standard primers, 2mM dNTPs plus one of the three currently available fluorescent dUTPs from ABI (R6G, R110 and TAMRA) at the manufacturer's recommended concentration. PCR products are then gel purified and resuspended in the same original volume of TE, and stored at 4°C.

Preparation of target DNA

Preparation of target DNA for use with end-labelled probes

Target sequences are amplified as usual except that biotinylated primers are used (1ng/µl), and up to four segments can be multiplexed in microtitre plate wells if required. No further purification is necessary.

Preparation of internally-labelled target DNA

For internal labelling, reactions are performed separately and include the appropriate modified dUTP and biotinylated primers. In this case 10µl reactions are sufficient.

Chemical cleavage of mismatch

With end-labelled probe

1) Add 5µl of each end-labelled purified probe to 20µl of the multiplexed biotinylated targets in hybridisation buffer (0.3M NaCl, 100mM Tris-HCl pH 8.0 final) in the microtitre plate. Incubate at 95°C for 5 min followed by 65°C for 60 min.

2) Prepare the streptavidin-magnetic beads (either Dynabeads or Promega) by washing twice with 2x binding buffer (2M NaCl, 0.4% Tween-20, 10mM Tris-HCl pH 8.0, 0.1mM EDTA), and then resuspending in the same volume of 2x binding buffer as the hybridisation reactions to which they are to be added. Add an equal volume of magnetic beads to each reaction and leave at room temperature for 15 min. Remove supernatant while on the magnet.

3) Add 20μl of hydroxylamine solution (4M hydroxylamine hydrochloride, 2.3M diethylamine) or osmium solution (0.4% osmium tetroxide, 2% pyridine), and incubate for 2 hr at 37°C and for 15 min at room temperature, respectively. Work inside fume hood.

4) Remove supernatant on magnet, resuspend beads in 5μl of the piperidine/dye solution (2M piperidine in formamide/dextran blue dye) and incubate at 90°C for 30 min.

5) Snap-chill on ice/water bath before loading supernatant on a 6% polyacrylamide/urea gel on the ABI 373 DNA sequencer.

With internally-labelled products

1) Combine 1μl of each of the three internally labelled probes with 1μl of each of the corresponding internally labelled biotinylated targets in one tube or microtitre well, and dilute to a final volume of 30μl in hybridisation buffer (as above).

2) After hybridisation add 30μl of magnetic beads (ie 10μl magnetic beads, washed and resuspended in 30μl of 2x binding buffer). Proceed as above.

Analysis of gels

The ABI 672 Genescan analysis software is used. Molecular weight markers (ABI Genescan 2500-ROX) are loaded in separate tracks and the sizes of fragments are estimated from the gel image. Adjustment of the gel contrast can help reveal fainter bands. Precise sizing is irrelevant as only the approximate location is required for subsequent sequencing of suspected mutations.

Comments

End-labelling is achieved by using HPLC-purified fluorescence-labelled primers in the PCR. The primers are stable for a long time if stored at 4°C. Currently five dyes are available (FAM, HEX, TET, TAMRA and ROX) although only four can be used simultaneously. If PCR conditions are optimised, then the PCR can be multiplexed as well as the mismatch reactions. However, only one dye is incorporated per single strand of

DNA, and since this is always at the 5' end, biotinylated primers cannot be used for the same reaction.

The fluorescence-labelled dUTPs that have recently become available allow 5' biotin primers to be used in the synthesis of strands that become internally labelled at variable positions with the fluorophores used in each reaction. This creates for almost every mutation two distinct opportunities to cause mismatch cleavage. Furthermore, as label is incorporated at random sites along the DNA, both cleavage bands are visible and add up to the full length probe, thereby confirming the true nature of any particular band, and allowing more accurate sizing. However, target DNA cannot be amplified in multiplex since different labels must be incorporated in separate amplification reactions. This is not a serious disadvantage as the major time-saving is to be had in multiplexing the mismatch reactions.

Only three fluorescent dUTPs are currently available, limiting the number of reactions that can be multiplexed.

Osmium tetroxide appears to react with the dUTP to produce a smeary band at around 170nt that could obscure a mismatch band of a similar size.

References

Cotton, R.G.H., Rodrigues, N.R., and Campbell, R.D. Reactivity of cytosine and thymine in single-base-pair mismatches with hydroxylamine and osmium tetroxide and its application to the study of mutations. *Proc. Natl. Acad. Sci. USA.* **85**: 4397-4401, 1988.

Haris, I.I., Green, P.M., Bentley, D.R., and Giannelli, F. Mutation detection by fluorescent chemical cleavage: application to haemophilia B. *PCR Meth. Applic.* **3**: 268-271, 1994.

Montandon, A.J., Green, P.M., Giannelli, F., and Bentley, D.R. Direct detection of point mutations by mismatch analysis: application to haemophilia B. *Nucl. Acids Res.* **17**: 3347-3358, 1989.

Naylor, J.A., Green, P.M., Montandon, A.J., Rizza, C.R., and Giannelli, F. Detection of three novel mutations in two haemophilia A patients by rapidly screening whole essential regions of the factor VIII gene. *Lancet* **337**: 635-639, 1991.

Roberts, R.G., Bobrow, M., and Bentley, D.R. Point mutations in the dystrophin gene. *Proc. Natl. Acad. Sci. USA.* **89**: 2331-2335, 1992.

Rowley, G, Saad, S., Giannelli, F. and Green, P.M. Ultrarapid mutation detection by multiplex, solid phase chemical cleavage. *Genomics*, in press 1995.

# CHAPTER • 13

EMC
ENZYME MISMATCH CLEAVAGE

Rima Youil* and Richard G.H. Cotton
*Communicating author

The enzyme mismatch cleavage (EMC) method for mutation detection (Youil et al., 1993; 1995) relies on a resolvase, T4 endonuclease VII, to detect mismatches and heteroduplex loops formed after mixing, melting, and annealing radioactively end-labelled normal DNA with unlabelled mutant DNA (or vice versa). The enzyme cleaves heteroduplex molecules within 6 bases 3' of a mismatch or a loop, allowing localization of the mutation. Separation of the cleavage products is performed on an 8% acrylamide-urea gel and the cleavage products are detected by autoradiography.

Protocol

PCR amplification

End-labelling

Heteroduplex reaction

T4 endonuclease VII cleavage reaction

Resolution of cleavage products

PCR amplification

Normal and mutant DNA (genomic, tumour, plasmid, or cDNA) are each amplified separately using the appropriate primers and PCR conditions. Purify the correct PCR product and determine the DNA concentration.

End-labelling

One hundred ng of the wild-type DNA is end-labelled with 1μl of 6,000Ci/mmol γ^{32}P-ATP and 10 units of 5'-polynucleotide kinase in a total volume of 15μl. Incubate for 45 min at 37°C. Precipitate using sodium acetate/ethanol. Centrifuge for 10 min before discarding the supernatant. From this point on check counts in the pellet using a hand held counter after each precipitation and wash step. Wash the pellet twice with 500μl 70% ice-cold ethanol (centrifuge for 1 min at each wash) to remove any free label. Air dry the pellet briefly at 37°C, and resuspend thoroughly in 10μl dH$_2$0. This should be enough probe for about four homoduplex and four heteroduplex reactions.

Heteroduplex reaction

Denaturation

1) Add 4μl of end-labelled wild type DNA (~40ng) to every 200-250ng of unlabelled mutant DNA (to give 500-1000cps). This will be enough for four reactions (eg 0, 250, 500 and 1000U T4 endonuclease VII).

2) Add 4μl of end-labelled wild-type DNA to 200-250 ng of unlabelled wild-type DNA. This will constitute the homoduplex control.

3) Resuspend each mixture to a total volume of 10μl with dH$_2$0 and add a further 10μl of 2x annealing buffer (1.2M NaCl, 12mM Tris-HCl pH 7.5, 14mM MgCl$_2$).

4) Boil the samples for 5 min.

Annealing

1) After boiling, immediately place the tubes on a heating block set at 65°C for 1 hr. Incubate for a further 60 min at room temperature before precipitating with sodium acetate/ethanol.

2) Wash the pellet with 500μl 70% ice-cold ethanol and resuspend the dried pellet thoroughly in 10μl dH$_2$0.

T4 endonuclease VII cleavage reaction

1) Add 5μl of 10x T4 endonuclease VII reaction buffer to 39μl of dH$_2$0 (to give a final reaction concentration of 50mM Tris-HCl pH 8.0, 10mM MgCl$_2$, 10mM dithiothreitol, 100mg/ml bovine serum albumin).

2) Add 5μl of the end-labelled duplex DNA (~50-100 cps).

3) Add 1μl of T4 endonuclease VII (250, 500 and 1000U/μl) or 1μl enzyme dilution buffer (10mM Tris-HCl pH 8.0, 0.1mM glutathione, 100μg/ml bovine serum albumin, 50% glycerol). The enzyme was kindly provided by Dr Böerries Kemper (University of Cologne,

Germany), and by Applied Technology Genetics Corporation (Malvern, PA, USA).

4) Incubate for 45-60 min at 37°C.

Stop

Sodium acetate/ethanol precipitate digestion products. Wash once with 70% ice-cold ethanol and resuspend the dried pellet in 6μl formamide gel loading dye, containing xylene cyanol and bromophenol blue. Vortex thoroughly.

Resolution of cleavage products

Loading

1) Preheat the 8% acrylamide-urea gel to 50°C.
2) Heat the samples at 100°C for 3 min and load immediately onto the gel.
3) Load ~150ng of end-labelled phiX 174 RF/*Hae*III digest size marker in one lane of the gel.
4) Run the gel till the bromophenol blue dye reaches the bottom of the gel.
5) Fix the gel in 10% methanol/10% acetic acid for 10 min.
6) Dry the gel down under vacuum.
7) Set up an autoradiographic exposure of the gel using intensifying screens. Set at -70°C overnight.
8) Develop the autoradiogram and determine the size of any cleavage bands detected by comparing to the phiX/*Hae*III size marker.
9) Sequence to define the mutation.

Comments

This mutation detection method relies on efficient formation of heteroduplexes. Both probe and unlabelled DNA must be completely denatured, followed by a slow reannealing reaction, to enable efficient formation of heteroduplexes. In our hands boiling probe and unlabelled DNA in 2x annealing buffer for 5 min, followed by incubation at 65°C for an hour and room temperature incubation for 60 min ensures adequate heteroduplex formation. A low heteroduplex yield may result in high background following enzyme cleavage.

T4 endonuclease VII produces a background cleavage pattern that is characteristic for the particular DNA fragment being scanned. It is therefore important to contain a homoduplex control DNA sample to determine the cleavage pattern and to compare each test sample with the control lane following enzyme digestion.

We have also noticed that a heavy background may be due to problems with the reaction buffer. The background may be reduced by adding freshly prepared dithiothreitol at a concentration of 100mM to a 10x reaction buffer.

The background may be further reduced by performing EMC on a solid support. By using biotinylated primers to amplify mutant target DNA, one can immobilise the duplex molecules on streptavidin-coated paramagnetic beads (e.g. Dynabeads, Dynal AS) in order to remove radiolabelled normal strands that have not hybridized and formed heteroduplex molecules with mutant DNA (Babon et al. in press).

A complete loss of DNA as shown by a clear lane in the test after autoradiography indicates that too much enzyme has been added to the EMC reaction such that the enzyme has completely degraded the DNA (unless the pellet was lost during precipitation or wash steps). The zero enzyme lane should show a strong band at the position of the intact fragment being scanned. If the 0 enzyme lane is very faint, this indicates that too little probe has been added to the heteroduplex reaction. In this situation it is best to repeat the heteroduplex reaction with freshly prepared probe that is adequately end-labelled, and to add more probe to the unlabelled DNA when forming heteroduplexes. If the 0 enzyme lane is strong but there is still overdigestion, then repeat this digestion using half or one fourth the enzyme concentrations originally tried. Aim for a digestion that leaves some of the probe intact.

In order to determine which end of the DNA fragment a mutation is closest to, heteroduplexes may be prepared where one or the other mutant strand is labelled with a biotin tag. This requires that heteroduplexes are prepared and assayed in two separate reactions. Alternatively, two primers each with a different fluorescence tag can be used to amplify the mutant DNA. By analysing the products on an automated fluorescence DNA sequencer it is possible to determine which end of the DNA fragment contains the mutation.

References

Babon, J.J., Youil, R., and Cotton, R.G.H. Improved strategy for mutation detection - A modification to the enzyme mismatch cleavage method. *Nucl. Acids Res.* in press.

Youil, R., Kemper, B. W., and Cotton, R.G.H. 1993 *Am. J. Hum. Genet.* Vol 53, Number 3 Supplement, Abstract No. 1257.

Youil, R., Kemper, B. W., and Cotton, R.G.H. Screening for mutations by enzyme mismatch cleavage with T4 endonuclease VII. *Proc. Natl. Acad. Sci. USA* 92: 87-91, 1995.

MREC
MISMATCH REPAIR ENZYME CLEAVAGE
A-Lien Lu-Chang

The E. coli MutY protein has been shown by Tsai-Wu et al. (1992) to have both DNA glycosylase and apurinic/apyrimidinic (AP) endonuclease activities. The DNA glycosylase activity removes the adenine bases from A/G and A/C mismatches and the AP endonuclease activity cleaves the first phosphodiester bond 3' to the AP site. Apparent dissociation constants are 5.3, 15, and 370nM for A/G-, A/C-, and C/G-containing DNA, respectively (Lu et al., 1995).

A novel mismatch repair enzyme cleavage (MREC) method has been developed to identify mutations using mismatch repair enzymes (Lu and Hsu, 1991). The specific binding and nicking of DNA fragments containing an A/G or A/C mismatch by MutY allowed the development of a powerful method for the detection and characterization of point mutations in the human genome. MutY can detect both A/T - C/G transversions and G/C - A/T transitions.

Other mismatch-recognizing enzymes can also be applied in the MREC method (Hsu et al., 1994). Mammalian thymine glycosylase removes thymines from T/G-, T/C-, and T/T-containing DNA (Neddermann and Jiricny, 1993) and is useful for detecting G/C - A/T transitions as well as A/T - C/G and T/A - A/T transversions. The all-type endonuclease or topoisomerase I from human or calf thymus (Yeh et al., 1994) can recognize all eight mismatches and can be used to screen any mutations. Sensitivity, reliability, and precise localization of the mutation site are the advantages of the MREC method.

The assay strategy is illustrated in Figure 1. Genomic DNA is isolated from normal and tumor cells, and the sequences of interest are amplified by PCR. The PCR products are denatured and renatured to form heteroduplexes as well as the original homoduplexes. Two types of

1. Isolation of genomic DNA
2. Amplify target DNA
3. Heteroduplex formation
4. Cut with mismatch repair enzymes
5. Analyze with sequencing gel

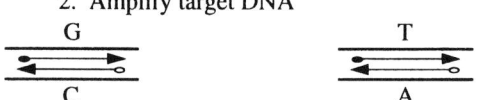

FIGURE 1. Detection of a mutation by the MREC method. Specific regions up to 1kb in length can be analyzed after amplification by PCR using primers labelled with two different fluorescent dyes.

heteroduplexes with base mismatches can form, for example, A/G- and C/T- containing DNA. These hybrids are subjected to cleavage with mismatch repair enzymes and then analyzed by denaturing gel electrophoresis.

Protocol

Amplification of genomic DNA

Heteroduplex formation

Labelling of DNA substrates

Purification of mismatch repair enzymes

MutY nicking assay

MutY binding assay

Thymine glycosylase assay

All-type endonuclease assay

Amplification of genomic DNA

1) About 10^7 cells (or 0.2-0.3g of tissue) is dissolved and digested for 1-2 hr in 1ml of SDS Tris-HCl pH 7.4, containing 200μg/ml proteinase K and 100μg/ml RNase A.

2) Genomic DNA is purified by two phenol-chloroform (1:1) extractions and precipitated with 2 vol. of ethanol.

3) Following two washes with 70% ethanol, the DNA is dissolved in TE buffer, pH 7.6.
4) Perform the PCR with synthetic primers for 30 cycles.
5) A second PCR amplification may be performed with primers 5' end-labelled with ^{32}P or with fluorescent dyes.
6) Purify the PCR products from an agarose or polyacrylamide gel.

Heteroduplex formation

1) Mix two PCR products (one from normal and one from mutant cells) or two oligonucleotides in the hybridization buffer.
2) Heat the DNA mixture at 90°C for 2 min and cool gradually over 30 min. Heteroduplexes, as well as the original homoduplexes, will be generated.

Labelling of DNA substrates

3'-end labelling of oligonucleotides

The Oligo-1 A/G duplex is a substrate for MutY. Oligo-2 C/G duplex is not a substrate and should also be used as a negative control for MutY nicking and binding.

1) To a microcentrifuge tube add the following:

Oligo-1 or -2	1µl
5x labelling buffer	3µl
[α-^{32}P] dCTP at 3,000Ci/mmol	5µl
Sterile dH$_2$O	5.5µl
Klenow fragment (5U/µl)	0.5µl
Total	15µl

2) Incubate the reaction for 30 min at room temperature.
3) Then add 1µl of 0.5M EDTA and 34µl of TE buffer to stop the reaction.

5' labelling of DNA using polynucleotide kinase

Oligonucleotides, PCR primers, or purified PCR products can be labelled by polynucleotide kinase with incorporation of γ-^{32}P-ATP at 3,000 Ci/mmol (Maniatis et al., 1982).

Removal of free nucleotides using Quick-spin G-25 or G-50 column

After labelling, remove free nucleotides by a Quick-spin G-25 or G-50 column, following the manual of Boehringer Mannheim, and check the incorporation by acid precipitation.

Purification of mismatch repair enzymes

1) E. coli MutY was purified according to the procedure described by Tsai-Wu et al. (1992). MutY protein is also available from Trevigen, Inc. Gaithersburg, Maryland, USA (Tel. 800-873-8443)

2) Mammalian thymine glycosylase, all-type endonuclease and topoisomerase I were purified according to published procedures (Neddermann and Jiricny, 1993; Yeh et al., 1994).

MutY nicking assay

The nicking activity of MutY is the combined action of the glycosylase and AP endonuclease activities. MutY nicks on the A-containing strand at the first phosphodiester bond 3' to the mismatched A.

MutY protein should be diluted 50-fold in the Storage/Dilution buffer.

1) For each reaction add the following to a microcentrifuge tube.

3' end-labelled 20-mer DNA (1.8fmol)	1µl
10x MutY reaction buffer	1µl
Sterile dH$_2$O	7µl
MutY (1/50)	1µl
Total	10µl

   Assay both A/G- and C/G-containing DNA. A control incubation consisting of DNA only (no MutY protein) should also be run. Incubate all reactions at 37°C for 30 min.

2) Stop the reactions in a dry ice-ethanol bath and dry in a desiccator for 45 min.

3) Resuspend each sample in 3µl of sequencing dye. Heat samples at 90°C for 2 min.

4) Analyze the reaction products on a 14% polyacrylamide DNA sequencing gel which has been prerun for 30 min.

5) Run the gel at 2,000 volts until bromophenol blue has migrated approximately half way down the gel.

6) Remove the glass plates and transfer the gel onto a used X-ray film for support.

7) Cover the gel with plastic film and autoradiograph until the proper exposure is achieved. It takes 16 hr for 3,000 cpm of DNA. The nicked product migrates just below the bromophenol blue and the intact DNA migrates between xylene cyanol and bromophenol blue.

Comment: For longer DNA segments, use an 8% sequencing gel.

MutY binding assay

The binding of MutY to DNA substrates can be assayed by gel retardation.

1) For each reaction, add the following to a microcentrifuge tube.

3' end-labelled 20-mer DNA (1.8fmol)	1µl
10x MutY reaction buffer	2µl
10µg/ml poly (dI-dC)	2µl
Sterile dH$_2$O	14µl
MutY, diluted 1/50	1µl
Total	20µl

   Assay both A/G- and C/G-containing DNA. A control incubation consisting of DNA only (no MutY protein) should also be run. Incubate all reactions at 37°C for 30 min.

2) Stop the reactions by adding 1.5 µl of 50% glycerol.

3) Load the entire reaction mixture onto an 8% polyacrylamide non-denaturing gel in 50 mM Tris-borate pH 8.3, 1mM EDTA, prerun for more than 30 min. Also load into an adjacent well with 1x DNA dye in TE (do not add dye to reaction mixtures).

4) Run the gel at 10volts/cm until bromophenol blue has migrated more than half-way down the gel.

5) Remove the glass plates and transfer the gel onto a 3MM filter paper.

6) Dry the gel in a gel dryer for 45 min and autoradiograph until the proper exposure is achieved. The free DNA migrates below the bromophenol blue and the MutY-bound complex migrates at a position near xylene cyanol.

Comment: For longer DNA, use a 4% native polyacrylamide gel.

Thymine glycosylase assay

The thymine glycosylase activity can be assayed in a manner similar to the MutY nicking assay except that a different buffer is used. DNA containing a T/G mismatch (Oligo-3) is a substrate for thymine glycosylase.

1) ^{32}P-labelled DNA heteroduplexes are incubated with the enzyme in 20µl of M buffer for 3 hr at 37°C.

2) Add 1µl of piperidine to the reaction, incubate at 90°C for 30 min.

3) Lyophilize the samples for 45 min.

4) Add 20µl of water and lyophilize for 45 min.

5) Dissolve the sample in 3µl of sequencing dye.

6) Denature DNA at 90°C for 2 min.

7) Fractionate the products on an 8% or 14% polyacrylamide-8.3M urea sequencing gel.

8) Autoradiograph the gel.

All-type endonuclease assay

Nicking activity is assayed similar to the assay for thymine glycosylase (Yeh et al., 1994). An oligonucleotide duplex containing an A/A or A/G mismatch is the substrate.

1) Incubate heteroduplexes and control homoduplexes with the enzyme in **M** buffer for 3 hr at 37°C. No piperidine treatment is required.

2) After incubation, samples with 5' end-labelled DNA are treated with 400µg/ml of protease K at 50°C for 1hr followed by phenol and ether extractions (this step is not necessary for 3' end-labelled DNA).

3) Samples are lyophilized and dissolved in 3µl of sequencing dye.

4) DNA is denatured at 90°C for 2 min.

5) Fractionate the products on an 8% or 14% polyacrylamide-8.3M urea sequencing gel.

6) Autoradiograph the gel.

Reagents

E. coli MutY protein: 10µl containing 1.1µg protein in Storage/Dilution buffer. MutY should be diluted 50-fold and 10-fold for A/G- and A/C-containing DNA, respectively.

HeLa or calf thymus thymine glycosylase at 20µg/ml

HeLa or calf thymus all-type endonuclease or topoisomerase at 11µg/ml

Oligonucleotide-1 A/G duplex: 1pmol/µl in Hybridization buffer.

(1) 5'-CCGAGGAATTAGCCTTCTG-3'
3'-GCTCCTTAAGCGGAAGACG-5'

Oligonucleotide-2 C/G duplex: 1pmol/µl in Hybridization buffer.

(2) 5'-CCGAGGAATTCGCCTTCTG-3'
3'-GCTCCTTAAGCGGAAGACG-5'

Oligonucleotide-3 T/G duplex: 1pmol/µl in Hybridization buffer.

(3) 5'-CCGAGGAATTTGCCTTCTG-3'
3'-GCTCCTTAAGCGGAAGACG-5'

Storage/Dilution buffer: 20mM potassium phosphate pH 7.4, 1.5mM dithiothreitol, 0.1mM EDTA, 50mM KCl, 200µg/ml bovine serum albumin, 50% glycerol

10x Hybridization buffer: 70mM Tris-HCl pH 7.6, 70mM MgCl$_2$, 500mM NaCl

5x 3' end-labelling buffer: 250mM Tris-HCl pH 7.6, 25mM MgCl$_2$, 25mM ß-mercaptoethanol, 0.1mM dGTP, 0.1mM dTTP

10x 5' end-labelling buffer: 500mM Tris-HCl pH 7.6, 100mM MgCl$_2$, 50mM dithiothreitol, 1mM spermidine, 1mM EDTA

Klenow fragment of DNA polymerase I or polynucleotide kinase

α^{32}P-dCTP or γ^{32}P-ATP at 3,000 Ci/mmol

Quick-spin G-25 or G-50 columns (Boehringer Mannheim)

10x MutY reaction buffer: 200mM Tris-HCl pH 7.6, 800mM NaCl, 10mM dithiothreitol, 10mM EDTA, 29% (v/w) glycerol

10x M buffer: 200mM Tris-HCl (pH 7.6), 10mM dithiothreitol, 10mM EDTA, 0.1mM ZnCl$_2$, and 29% (w/v) glycerol

Poly (dI-dC), 10μg/ml

TE: 10mM Tris-HCl pH 7.6, 1mM EDTA

Sequencing dye: 90% formamide, 10mM EDTA, 0.1% xylene cyanol, 0.1% bromophenol blue

10x DNA dye: 60% glycerol, 50mM EDTA, 0.5% SDS, 0.05% xylene cyanol, 0.05% bromophenol blue

References

Hsu, I.-C., Yang, Q., Kahng, M. W., Xu, J.-F. Detection of DNA point mutation with DNA mismatch repair enzymes. *Carcinogenesis* 15: 1657-1662, 1994.

Lu, A-L., and Hsu, I.-C. Detection of single DNA base mutations with mismatch repair enzymes. *Genomics* 14: 249-255, 1991.

Lu, A.L., Tsai-Wu, J.J., and Cillo, J. DNA determinants and substrate specificities of *Escherichia coli* MutY. *J. Biol. Chem.* 270: 23582-23588, 1995.

Maniatis, T., Fritsch, E. F., and Sambrook, J. Molecular Cloning: A laboratory manual (Cold Spring Harbor Laboratory, Cold Spring Harbor, NY), 1982.

Neddermann, P. and Jiricny, J. The purification of a mismatch-specific thymine-DNA glycosylase from HeLa cells. *J. Biol. Chem.* 268: 21218-21224, 1993.

Tsai-Wu, J.-J., Liu, H.-F., and Lu, A-L. Escherichia coli MutY protein has both N-glycosylase and apurinic/ apyrimidinic (AP) endonuclease activities on A/C and A/G mispairs. *Proc. Natl. Acad. Sci. USA* 89: 8779-8783, 1992.

Yeh, Y.-C., Liu, H.-F., Ellis, C. A. and Lu, A.-L. Mammalian topoisomerase I has base mismatch nicking activity. *J. Biol. Chem.* 269: 15498-15504, 1994.

**DETECTION OF
KNOWN MUTATIONS**

SSO
GENETIC TYPING WITH SEQUENCE-SPECIFIC OLIGONUCLEOTIDES

Ulf Gyllensten* and Marie Allen
*Communicating author

One of the most common methods for distinguishing DNA sequence variants is based on hybridization of short sequence-specific oligonucleotides to PCR-amplified target sequences. The use of oligonucleotide hybridization to distinguish genetic variants predates PCR. However, hybridization of oligonucleotides to single-copy sequences in complex genomic DNA is exceedingly difficult, and the method therefore became widely applied only after large numbers of copies of specific fragments could be generated through PCR. A number of formats have been developed for genetic typing by oligonucleotide hybridization.

1) Forwards dot blots. The amplified DNA is attached to a membrane or other solid support and it is hybridized to probes in solution.

2) Reverse dot blots. The probes are attached to a solid support and hybridized to labelled PCR products in solution.

3) Liquid phase hybridization. Both PCR products and probes are in solution during hybridization. In the latter case, the product-probe complex may be separated from free product and probe by gel electrophoresis.

Although the last method is more cumbersome and less frequently used, it provides a very high sensitivity since solution hybridization is more efficient than hybridization between an immobilized DNA strand and a complementary molecule in solution.

Protocol

Design of probes

Labelling of probes

Hybridization and washing

Detection systems

Data processing and genotype calling

Design of probes

A correct design of hybridization probes is the basis for reliable discrimination between genetic variants. In selecting hybridization probes we make sure that the following criteria are met:

- The oligonucleotides are selected to have melting temperatures (T_m) of 42-48°C. A number of methods are available for estimating the approximate T_m of an oligonucleotide and these yield slightly different values. We usually employ the simple Wallace rule (2°C per A or T and 4°C per G or C).
- Avoid internally complementary sequences of more than three base pairs, and sequences that can form internal hairpins.
- Try to find the most destabilizing mismatch, by selecting the most suitable strand as a target. This is only possible if forward dot blots are used or if both strands are labelled in reverse dot blots.
- In the case of single nucleotide differences, position the mismatch one-third in from one end of the oligonucleotide, not in the middle.

A summary of the rules for designing oligonucleotide hybridization probes are given in Table 2 of Stoneking et al. (1991).

Labelling of probes

In forward dot blots, the probes were initially labelled with $\gamma^{32}P$-ATP. However, radioactive probes require extra safety precautions and the hybridization results are usually more difficult to interpret, due to the strong signal and concomitant prominent nonspecific hybridization. Currently, most probes are either labelled with 5' biotin for further binding to streptavidin-conjugated horse radish peroxidase (SA-HRP), or peroxidase can be coupled directly to an oligonucleotide. In addition, a number of fluorophores are now available for coupling to oligonucleotides, increasing the opportunities for highly sensitive and automated detection of hybridization.

Hybridization and washing

Forward dot-blots

After PCR, 5-20µl of amplification product is denatured in 200µl of 0.4M NaOH, 25mM EDTA for 20 min, and 50µl of this mixture is applied to a prewetted Biodyne B membrane (Pall BioSupport, East Hills, NY, USA), using a Dot-Blot Apparatus (BioRad). These filters do not require DNA immobilization by UV irradiation and they can be reused multiple times after stripping with 1% SDS in water at 80°C. After applying the PCR products, the membranes are prehybridized in 2x SSPE (1x SSPE is 0.18M NaCl, 10mM NaH_2PO_4, and 1mM EDTA pH 7.7) and 0.5% SDS at 40°C to remove excess PCR products. Hybridization is carried out for 10-30 min in a fresh aliquot of the same buffer, preheated to 42°C, with the addition of probe at a final concentration of 0.5-1pmol/µl. For probes labelled with biotin and peroxidase-based detection, SA-HRP can be added directly to the hybridization solution. Probes conjugated with peroxidase can be used at concentrations lower than biotinylated probes that require secondary binding of SA-HRP.

After hybridization the membranes are washed at a suitable temperature, depending on the T_m of the oligonucleotide probe. For a set of probes with a T_m of 42°C (calculated by the Wallace method) that we use to distinguish genital human papilloma virus types, we typically use 1x SSPE, 0.1% SDS for 20 min at 42 or 50°C (Ylitalo et al., 1995).

Detection systems

Hybridization of nonradioactively labelled probes can be conveniently detected via conversion of luminol by HRP, resulting in the emission of light of wavelengths 350-550nm (peak at 444nm). There are several different manufacturers of substrate for the detection reaction (e.g. ECL, Amersham). Probes that are directly conjugated with HRP usually result in a 2-3 times higher signal. However, such probes are more expensive to synthesize than biotinylated ones, and since they should not be frozen they may have a more limited shelf life. Also, by using biotinylated probes different enzymatic reactions can be used to detect hybridization. An example of the results of using the forward dot blot system and a series of oligonucleotide probes for genetic typing of genital human papilloma virus types from cervical smears is shown in the Fig. 1.

Data processing and genotype calling

The release of light can be detected by exposure of X-ray film (e.g. Kodak Xomat-S) or it can be measured in a luminometer. If a large number of probes are being used, and genotype calling is based on the hybridization pattern of combinations of probes, a computerized database of potential hybridization patterns will speed up the genotype calling and improve the accuracy. A number of software programs exist in which an observed hybridization pattern can be entered as a string or zeroes and ones. Hybridization patterns represented by such binary codes can be used to search a custom-designed database for matched patterns.

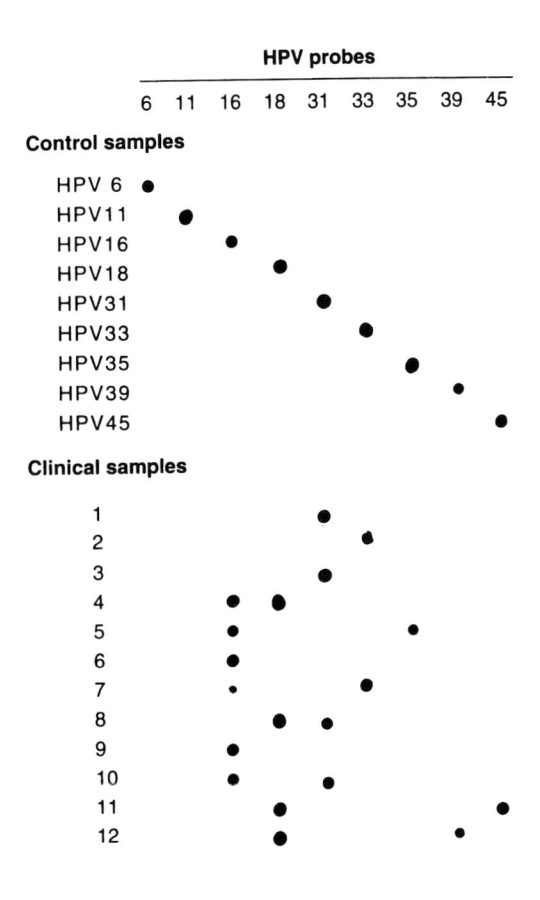

	\multicolumn{9}{c	}{HPV probes}
	6	11	16	18	31	33	35	39	45
Control samples									
HPV 6	●								
HPV11		●							
HPV16			●						
HPV18				●					
HPV31					●				
HPV33						●			
HPV35							●		
HPV39								●	
HPV45									●
Clinical samples									
1				●					
2				●					
3				●					
4			●	●					
5			●			●			
6			●						
7			●		●				
8			●	●					
9			●						
10			●	●					
11				●				●	
12				●		●			

FIGURE 1. Forward dot blot detection of strains of human papilloma virus. Amplification products were spotted on a Biodyne C membrane and hybridized with oligonucleotide probes, followed by peroxidase-mediated detection.

References

Stoneking, M., Hedgecock, D., Higuchi, R.G., Vigilant, L., and Erlich, H.A. Population variation of human mtDNA control region sequences by enzymatic amplification and sequence-specific oligonucleotide probes. *Am. J. Hum. Genet.* **48**: 370-382, 1991.

Ylitalo, N., Bergström, T., and Gyllensten, U. Detection of genital human papillomavirus by single tube nested PCR and type-specific oligonucleotide hybridization. *J. Clin. Micro.* **33**: 1822-1828, 1995.

PASA
PCR AMPLIFICATION OF SPECIFIC ALLELES
Steve S. Sommer

PASA is a generally applicable technique for the detection of known single-base substitutions or microdeletions/insertions. In this PCR-based technique, one of the PCR primers precisely matches one allelic variant of the target sequence, but it is mismatched to the other. When the mismatch occurs at or near the 3' end of the PCR primer, preferential amplification of the perfectly matched allele is obtained (Sommer et al., 1989). This technique is also known as the amplification refractory mutation system (ARMS; Newton et al., 1989), allele-specific PCR (ASPCR; Wu et al., 1989), and allele-specific amplification (ASA; Okayama et al., 1989).

PASA can generally detect a single copy of a mutant allele in the presence of 40 copies of the normal allele, and may also be used to perform haplotyping of nearby loci in the absence of relatives through double PASA. PASA shows promise for population screening because the technique is rapid, reproducible, inexpensive, nonisotopic, and amenable to automation.

Protocol

Primer design
PCR amplification
PASA optimization
Double PASA

Primer design

The principle is to design a primer that will preferentially amplify one allele over another. Specificity of amplification can be obtained if the oligonucleotide matches the desired allele, but is mismatched to the other allele at the 3' end of the oligonucleotide. The desired allele is readily amplified, while the mismatched allele is poorly amplified if at all. The poor amplification is a result of the mismatch between the DNA template and the oligonucleotide primer that prevents efficient 3' elongation by *Taq* polymerase. Primers are generally designed to have a "Wallace temperature" [(A+T)x2°C+(G+C)x4°C] of 48°-50°C. The primer sequence should (i) have a G+C content of about 50% if at all possible; (ii) should not have self-complementary sequences of 4 or more bp; and (iii) not have 4 or more nucleotide complementarity between its 3' end and that of the other PCR primer.

PCR amplification

PCR is performed using 30 cycles of 1 min denaturation at 94°C, 2 min annealing at 50°C, and 3 min extension at 72°C in a Perkin-Elmer Cetus automated thermal cycler. Reaction components include approximately 250ng of genomic DNA, 10mM Tris-HCl pH 8.3, 50mM KCl, 200µM of each deoxyribonucleotide, 0.5U of *Taq* polymerase, 1.5-4.5mM of $MgCl_2$ (see PASA optimization), and 0.05-1µM of each oligonucleotide (see PASA optimization) in a 25µl reaction volume.

PASA optimization

In the majority of cases, a standard magnesium titration (1.5, 2.5, 3.5, and 4.5mM) and oligonucleotide titration (1.0, 0.25, 0.1, and 0.05µM) are sufficient to provide conditions for a both robust and specific amplification. Often there is a wide range of acceptable magnesium and oligonucleotide concentrations. Ideally, magnesium and oligonucleotide concentrations are adjusted to also produce spurious amplification products that do not interfere with specific detection but provide an internal control for the technical success of the amplification when the specific band is absent.

Occasionally, further optimization is required for specificity, as listed below:

- Oligonucleotide concentration: Decreasing the oligonucleotide concentration to 0.05µM may increase specificity. However, below 0.025µM the amplification signal generally becomes weak.

- Magnesium concentration: Specificity can sometimes be achieved by lowering the magnesium concentration below 1.5mM. Adding EDTA is a simple way of decreasing the "effective" magnesium concentration without making a different PCR buffer/salt stock solution. Likewise, increasing magnesium concentrations above 4.5mM can produce non-interfering spurious amplification products that are useful as internal controls.

- DNA concentration: A 10-fold dilution of the standard genomic DNA concentration can increase specificity and still provide an adequate amplification signal. Diluting the template can also increase sensitivity and avoid problems caused by contamination of the DNA with any PCR inhibitors.

- Thermal cycler specificity: Once a PASA reaction has been optimized on a specific thermal cycler, care should be taken when doing the same reaction on another machine because of inconsistencies in the ramp times between temperatures among different thermal cyclers.

- Allele-nonspecific oligonucleotide: Occasionally, a given pair of primers will not provide specific amplification products. Surprisingly, replacement of the allele-nonspecific primer with another oligonucleotide at a new location will often provide specificity. Generally, primer pairs are chosen to yield products that range in size between 300 and 600 base pairs. However, this is not always possible as in the case of double PASA (see below). We have specifically amplified segments from 200 to 2700 bases pairs in length.

- Allele-specific oligonucleotide: Designing the allele-specific primer is critical. Selecting oligonucleotides with a lower Wallace T_m, around 42-44°C, and placing the mismatch at the 3' base will increase specificity. Occasionally a segment does not amplify well. Designing new primers using the other DNA strand for allele-specific mismatch may provide better amplification.

- Deoxyribonucleotide concentration: Decreasing the concentration of dNTPs to 25-50μM can prevent spurious amplification.

- Formamide and DMSO: Inclusion of formamide (typically 2-5%) can increase the signal strength and eliminate undesired spurious bands, especially in regions of high G +C content. In addition, allele specificity may be enhanced. Occasionally DMSO is effective when formamide is not.

- A second pair of primers: A primer set specific for another region of the genome can be added to the reaction to generate a constant band. This serves both as an internal control for the technical success of the PCR and to increase specificity by providing another substrate for the Taq polymerase.

- *Taq* polymerase: Decreasing the amount of enzyme in each reaction (0.2-0.3U/μl) can increase specificity, while the addition of more enzyme can create spurious bands to serve as internal controls.

- Source of the DNA template: Using a PCR product as the source of DNA template in a nested PASA can increase specificity, particularly if the region to be amplified is highly repetitive. Note, however, that

the concentration of DNA is critical and a 10^6-fold or greater dilution of the original PCR product may be required.
- Number of amplification cycles: Decreasing the number of PCR cycles may reduce detection of any amplification products of the mismatched allele. However, the number of cycles makes little difference in specificity.
- Annealing temperature: Raising the PCR annealing temperature can increase specificity. In our experience with optimization of over 65 PASA reactions, it has never been necessary to deviate from our standard cycling parameters (see above), except to increase the 3 min extension time at 72° for segments greater than 1.5 kb.

Double PASA

PASA can be adapted to provide haplotyping of an individual in the absence of relatives by utilizing pairs of allele-specific PCR primers. This "double PASA" differentially amplifies each haplotype. Four amplifications can distinguish the haplotypes produced by a pair of biallelic polymorphisms.

Double PASA is an important tool for haplotyping doubly heterozygous individuals because the physical linkage of alleles on a strand of DNA is necessary to determine the haplotype. Double PASA should be generally applicable for haplotyping, provided that the segment between the polymorphisms can be amplified at least to a moderate extent. If the polymorphic sites are separated by too great a distance to allow PCR amplification, double PASA can be combined with inverse PCR (Loh et al., 1989). By circularization, the genomic targets can be placed close enough together to allow inverse PCR.

References

Bottema, C.D.K., Sarkar, G., Cassady, J.D., Ii, S., Dutton, C.M., and Sommer, S.S. PCR amplification of specific alleles: A general method of rapidly detecting mutations, polymorphisms, and haplotypes. *Meth. Enzymol.* **218**: 388-402, 1993.

Bottema, C.D.K., and Sommer, S.S. PCR amplification of specific alleles: rapid detection of known mutations and polymorphisms. *Mut. Res.* **288**: 93-102, 1993.

Loh, E.Y., Elliott, J.F., Cwirla, S., Lanier, L.L., and Davis, M.M. Polymerase chain reaction with single-sided specificity: Analysis of T cell receptor delta chain. *Science* **243**: 217-220, 1989.

Newton, C.R., Graham, A., Heptinstall, L.E., Powell, S.J., Smith, J.C., and Markham, A.C. Analysis of any point mutation in DNA. The amplification refractory mutation system (ARMS). *Nucl. Acids Res.* **17**: 2503-2516, 1989.

Okayama, H., Curiel, D.T., Brantly, M.L., Holmes, M.D., and Crystal, R.D. Rapid nonradioactive detection of mutations in the human genome by allele-specific amplification. *J. Lab. Clin. Med.* **114**: 105-113, 1989.

Sarkar, G. and Sommer, S.S. Haplotyping by double PCR amplification of specific alleles. *BioTechniques* **10**: 436-440, 1991.

Sommer S.S., Cassady, J.D., and Sobell, J.L. A novel method for detecting point mutations or polymorphisms and its application to population screening for carriers of phenylketonuria. *Mayo Clin. Proc.* **64**: 1361-1372, 1989.

Sommer, S.S., Groszbach, A., and Bottema, C.D.K. PCR amplification of specific alleles (PASA) is a general method for rapidly detecting known single-base changes. *Biotechniques* **12**: 82-87, 1992.

Wu, D.Y., Ugozzoli, L., Pal, B.K., Wallace, R.B. Allele-specific amplification of β-globin genomic DNA for diagnosis of sickle cell anemia. *Proc. Natl. Acad. Sci. USA* **86**: 2757-2760, 1989.

SOLID-PHASE
MINISEQUENCING

Ann-Christine Syvänen

The solid-phase minisequencing method detects single nucleotide variations in amplified DNA. It is also a useful tool for quantitative PCR analysis. The method is based on extension of a detection primer that anneals immediately adjacent to a variable nucleotide position on an affinity-captured amplified template, using a DNA polymerase and a single, labelled nucleoside triphosphate (Fig. 1) (Syvänen et al., 1990; 1993)

Protocol

PCR amplification

Affinity-capture

Denaturation

The minisequencing reaction

Measurement

Interpretation

PCR amplification
Carry out the PCR amplification with one biotinylated primer at a concentration of 0.2μM and one unbiotinylated primer at 1μM. The amount of biotinylated primer is reduced so as not to exceed the biotin-binding capacity (2-5pmol) of the streptavidin-coated wells in the affinity-capture step.

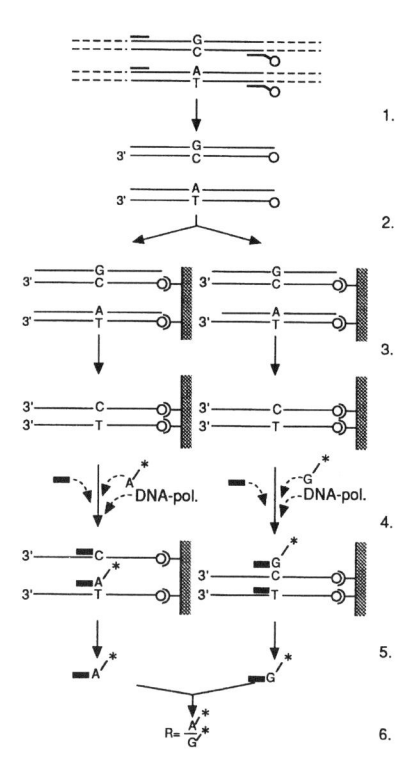

FIGURE1. Principle of the solid-phase minisequencing method illustrated by detection of an A/G polymorphism. Step 1: PCR amplification with one biotinylated and one unbiotinylated primer. Step 2: Affinity capture on an avidin- or streptavidin-coated solid support. Step 3: Washing and denaturation. Step 4: The mini sequencing reaction. Step 5: Measurement of the incorporated label. Step 6: Interpretation of the result.

Affinity capture

1) Transfer two 10µl aliquots of the PCR products, or four 10µl aliquots for duplicate assays, to streptavidin-coated microtiter wells (eg Combiplate 8, Streptavidin-coated, Labsystems).

2) Add to the wells 40µl of 20mM sodium phosphate buffer pH 7.5, 100mM NaCl, 0.1% Tween 20.

3) Seal the wells with a sticker and incubate at 37°C for 1.5 hr with gentle shaking.

4) Wash the wells 3 times manually with 200µl of 40mM Tris-HCl pH 8.8, 1mM EDTA, 50mM NaCl, 0.1% Tween 20 or 5 times using an automatic microtiter plate washer.

Denaturation

1) Add 100µl of 50mM NaOH to each well, and incubate at 20°C for 3 min.

2) Wash as above.

The minisequencing reaction

1) Prepare two master mixtures for detecting the nucleotides at the variable site. For each reaction, combine:

    5µl of 10x DNA polymerase buffer

    2µl of 5µM detection primer (a 20-mer designed to anneal immediately adjacent to the variable site)

    0.1µCi (usually 0.1µl) of one ^{3}H-labelled dNTP complementary to the nucleotide to be detected ([^{3}H]dATP, TRK 625; [^{3}H]dCTP, TRK 576; [^{3}H]dGTP, TRK 627; [^{3}H]dTTP, TR 633, Amersham)

    0.1U DNA polymerase (*Taq* or Dynazyme™, Finnzymes) and distilled H_2O to 50µl.

2) Add 50µl of reaction mixture per well.
3) Incubate at 50°C for 10 min for the primer extension step.
4) Wash as above.

Measurement

1) To release the extended detection primer, add 60µl of 50mM NaOH to each well and incubate at 20°C for 3 min.
2) Measure released ^{3}H in a scintillation counter.

Interpretation

In a homozygous sample a positive signal is generated in one of the minisequencing reactions, and in a heterozygous sample signals are generated in both reactions. Positive signals are usually 1000cpm or higher, depending on the efficiency of the PCR, and negative signals or background values are below 100cpm. Calculate the ratio between the two ^{3}H-labelled dNTPs incorporated in the reactions for each sample. The ratio is >10 or <0.1 for samples from individuals homozygous for either of the nucleotides. For samples from heterozygous individuals the ratio is between 0.5 and 2.0, depending on the specific activities of the ^{3}H-labelled dNTPs used.

Comments

- It is important that the PCR amplification is efficient if ^{3}H-labelled dNTPs are used as detectable groups as they have a low specific activity. (Ten µl of the PCR product should be clearly visible on an agarose gel by staining with ethidium bromide.)

- Other avidin- or streptavidin-coated solid supports with higher biotin-binding capacity than microtiter plate wells, such as microparticles, can also be used (Syvänen et al., 1990; 1992a; 1992b).
- The excess of dNTPs from PCR must be completely removed by the washing procedure in order to obtain specific minisequencing reactions.
- The same minisequencing reaction conditions are applicable irrespectively of the sequence of the detection step primer.
- Other DNA polymerases (eg T7 DNA polymerase) also perform well (Syvänen et al., 1990; 1992a).
- By using streptavidin-coated microtiter plates made of scintillating polystyrene, the final washing, denaturation and transfer of the released primer can be omitted, but a scintillation counter for microtiter plates is needed (Ihalainen et al., 1994).
- dNTPs labelled with other isotopes (Syvänen et al., 1990) or with colorimetrically detectable haptens (Harju et al., 1994; Livak and Hainer, 1994) have also be used.

Quantitative analysis

The ratio between the labels incorporated in the minisequencing reaction reflects the ratio between two sequences also when these are present in any other ratio than that in samples from homozygous (allele ratio 2:0) or heterozygous (allele ratio 1:1) subjects. Since the two sequences are essentially identical, they are amplified with equal efficiency during PCR. The minisequencing method thus allows convenient and accurate determination of the relative amounts of two sequences that differ from each other at a single nucleotide and are present as a mixture in a sample. Such quantitative PCR analysis has been used to determine the population frequencies of mutant alleles (Syvänen et al., 1992b) and polymorphisms (Syvänen et al., 1993) from large pooled DNA samples, the proportion of heteroplasmic mutations of the mitochondrial DNA (Suomalainen et al., 1992) and mutant blast cells in bone marrow samples (Syvänen et al., 1992a).

To determine the absolute amount of a sequence present in a sample, a known amount of an internal standard, differing from the target sequence at one nucleotide is added to the sample before the PCR amplification. The ratio obtained in the minisequencing assay, reflecting the ratio between the two sequences, allows calculation of the original amount of target sequence (Syvänen and Peltonen, 1994). This approach has been used to determine mRNA levels (Ikonen et al., 1992; Suomalainen et al., 1993) and the copy number of genes (Laan et al., 1995).

In addition to the relative amount of two sequences present in a sample, the ratio of signals obtained in a minisequencing assay is also affected by the specific activities of the ^{3}H-dNTPs used. If the sequence immediately next to the variable site contains one (or more) nucleotides identical to the one at the variable site, the primer will be extended by two (or more) dNTPs, which obviously also affects the obtained ratio. Both these factors can easily be corrected for. Alternatively, a standard curve can be constructed by mixing the two sequences in known ratios. In addition to correcting for the factors mentioned above, a standard curve will correct for possible misincorporation of small amounts of a dNTP by the DNA polymerase, which may affect the result when a sequence present as a small minority of a sample is to be quantified. Because of the high specificity of the primer extension reaction catalyzed by a DNA polymerase, the solid-phase minisequencing method allows quantification of one sequence variant, representing less than 0.1% of the target sequences (Syvänen et al., 1992a).

References

Harju, L., Weber, T., Alexandrova, L., Lukin, M., Ranki, M., Jalanko, A. Colorimetric solid-phase minisequencing assay illustrated by detection of α_1-antitrypsin Z mutation. *Clin. Chem.* 39: 2282-2287, 1993.

Ihalainen, J., Siitari, H., Laine, S., Syvänen, A.-C., Palotie, A. Towards automatic detection of point mutations: use of scintillating microtitration plates in solid phase minisequencing. *BioTechniques* 16: 938-943, 1994.

Ikonen, E., Manninen, T., Peltonen, L., Syvänen, A.-C. Quantitative determination of rare mRNA species by PCR and solid-phase minisequencing. *PCR Meth. Applic.* 1: 234-240, 1992.

Laan, M., Grön-Virta, K., Salo, A., Aula, P., Peltonen, L., Palotie, A., Syvänen, A.-C. Solid-phase minisequencing confirmed by FISH analysis in determination of gene copy number. *Hum. Genet.* 96: 275-280, 1995.

Livak, K.J, Hainer, J.W. A microtiter plate assay for determining apolipoprotein E genotype and discovery of a rare allele. *Hum. Mut.* 3: 379-385, 1994.

Suomalainen, A., Kollmann, P., Octave, J.-N., Söderlund, H., Syvänen, A.-C. Quantification of mitochondrial DNA carrying the tRNA$_{8344}^{Lys}$ point mutation in myoclonous epilepsy and ragged-red fiber disease. *Eur. J. Hum. Genet.* 1: 88-95, 1992.

Suomalainen, A., Majander, A., Pihko, H., Peltonen, L., Syvänen, A.-C. Quantification of tRNA$_{3243}^{Leu}$ point mutation of mitochondrial DNA in MELAS patients and its effects on mitochondrial transcription. *Hum. Mol. Genet.* 2: 525-534, 1993.

Syvänen, A.-C., Aalto-Setälä, K., Harju, L., Kontula, K., Söderlund, H. A primer-guided nucleotide incorporation assay in the genotyping of apolipoprotein E. *Genomics* 8: 684-692, 1990.

Syvänen, A.-C., Söderlund, H., Laaksonen, E., Bengtström, M., Turunen, M., Palotie, A. N-ras gene mutations in acute myeloid leukemia: accurate detection by solid-phase minisequencing. *Int. J. Cancer* 50: 713-718, 1992a.

Syvänen, A.-C., Ikonen, E., Manninen, T., Bengtström, M., Söderlund, H., Aula, P., Peltonen, L. Convenient and quantitative determination of the frequency of a mutant allele using solid-phase minisequencing: Application to aspartylglucosaminuria in Finland. *Genomics* **12**: 684-692, 1992b.

Syvänen, A.-C., Sajantila, A., Lukka, M. Identification of individuals by analysis of biallelic DNA-markers using PCR and solid-phase minisequencing. *Am. J. Hum. Genet.* **52**: 46-59, 1993.

Syvänen, A.-C., Peltonen, L. Accurate quantitation of rare mRNA species by polymerase chain reaction and solid-phase minisequencing. In Cell Biology: A Laboratory Handbook (ed. Celis, J) Academic Press Inc., pp 488-496, 1994.

MULTIPLEX SOLID-PHASE FLUORESCENT PRIMER EXTENSION

John M. Shumaker, Andres Metspalu*, and C. Thomas Caskey

*Communicating author

Solid phase primer extension is a mutation analysis and comparative sequencing method that uses allele-specific oligonucleotides for primer extension by T7 DNA polymerase to scan simultaneously for several previously known mutations in genes of known sequence. Three steps are required: a) template annealing to a set of oligonucleotides, b) oligonucleotide extension by incorporation of one fluorescent ddNTP, using T7 DNA polymerase, and c) separation of oligonucleotides by length in a fluorescent sequencer.

In this method a biotinylated PCR amplification is performed of the regions of interest, followed by isolation and purification of single stranded template via magnetic separation. PCR primers can be optimized for multiplex amplification. Individual extension primers are designed to vary in length, and anneal immediately upstream of each potential mutation site. A fluorescent ddNTP addition is performed using T7 DNA polymerase to distinguish between the mutant and wild type sequence. The products are purified from unincorporated ddNTPs, eluted with formamide, and analyzed on an automated fluorescent DNA sequencer. This method provides a parallel analysis format since oligonucleotide length is used as a tag for the site of the mutation on the DNA sequence, and the color of the peak in the chromatogram identifies the nucleotide at the potential site of mutation.

Protocol

PCR amplification

Immobilization of the PCR product, strand separation, and single strand recovery

Primer extension reactions

Analysis using a fluorescent DNA sequencer

Interpretation of the results

PCR amplification

Set up a 50µl multiplex PCR reaction by taking 5µl 10x PCR buffer, 5µl 2mM dNTP mix, 3µl DNA template (100 ng/ml), 15pmol of biotinylated primer, and 40pmol of nonbiotinylated primer for each region of interest and 0.5U of Ampli*Taq* (Cetus Corporation). Adjust the volume to 50µl with H_2O.

Immobilization of the PCR product, strand separation, and single strand recovery

Take 20µl of a 50µl amplification reaction to capture on 35mg of Dynal M-280 streptavidin-coated magnetic beads (Dynal A.S., Oslo, Norway). Follow the instructions provided by the manufacturer to isolate both the biotinylated and the nonbiotinylated strands.

Primer extension reactions

Extension primers are annealed are annealed immediately upstream of the sites of the potential mutations on the DNA template. For each annealing reaction, take 7µl (1pmole of template) of the ssDNA template bound to magnetic beads and mix with 2µl of 5x T7 buffer, and 1µl of primer mix (containing 3pmole of each primer). Heat the reaction mix to 80°C and cool to room temperature. Add 1µl 0.1M DDT, 3µl of *Taq* Dye Terminators ddNTPs (ABI, Foster City, CA) (final concentrations 0.018µM fluorophore-modified A and G terminators, and 0.075µM modified C and T terminators), and 2µl diluted (1:8, 1.5U/µl) Sequenase Version 2.0 T7 DNA polymerase (United States Biochemical, Cleveland, OH) to the annealed mixture. Stop the reaction after 1 min by adding 100µl 2 x SSPE/10% ethanol (washing buffer). Collect the magnetic beads, wash twice with 100µl of washing buffer, and elute the extended primers in 6µl of formamide.

Analysis using a fluorescent DNA sequencer

The primer extension products are analyzed by electrophoresis in a 16% (1:19) denaturing (8M urea) polyacrylamide gel, using an ABI Model 373 DNA sequencer (Foster City, CA). One µl of the reaction product is loaded in each lane of the gel.

Interpretation of the results

The electrophoresis band patter is easily interpreted as the oligonucleotide length identifies the mutation site and the fluorescence emission of the modified ddNTP identifies the site of the mutation analyzed. Mutations are revealed as a color change from the wild type extension products. The efficiency of incorporation of ddNTPs by T7 DNA polymerase provides an excellent signal to noise ratio in this method. For example, heterozygous positions are usually clearly identified by a band with a combination of the color of the normal and the mutant sequence variant. Moreover, in principle this method should detect all types of mutations except for repeated sequences longer than the length of the primer, as well as any polymorphisms within a given sequence.

The number of mutations that can be analyzed by this method is limited only by the possibility to produce oligonucleotides with distinct migration rates in a PAGE assay. Using an ABI 373 DNA sequencer, 36 individual DNA samples can be analyzed in a single run, each for 30-50 different mutations.

References

Livak, K.J. and Hainer, J.W. A microtiter plate assay for determining apolipoprotein E genotype and discovery of a rare allele. *Hum Mutat.* **3**: 379-385, 1994.

Shumaker, J.M., Metspalu, A. and Caskey, C.T. Mutation detection by solid phase primer extension. *Hum. Mutat.* in press, 1995.

Sokolov, B.P. Primer extension technique for the detection of single nucleotide in genomic DNA. *Nucl.Acids Res.* **18**: 3871, 1989.

Syvänen, A.C., Aalto-Setala, K., Jarju, L., Kontula, K., and Söderlund, H. A primer-guided nucleotide incorporation assay in the genotyping of apoliproprotein E. *Genomics* **8**: 684-692, 1990.

OLA
DUAL-COLOR OLIGONUCLEOTIDE LIGATION ASSAY

Martina Samiotaki, Marek Kwiatkowski, and Ulf Landegren*

*Communicating author

Oligonucleotide ligation-assisted analysis of DNA sequences is characterized by 1) highly specific identification of DNA sequences in complex DNA samples, 2) accurate distinction among known sequence variants, and 3) easy adaptation to standardised, automated assay formats. Here we present a protocol for dual-color analysis of allelic sequence variants in amplified DNA sequences (Samiotaki et al., 1994). The steps of the assay are: target sequence amplification by PCR, introduction of allele-specific ligation probes, and capture and detection of ligated probes by time-resolved fluorometry (Fig. 1).

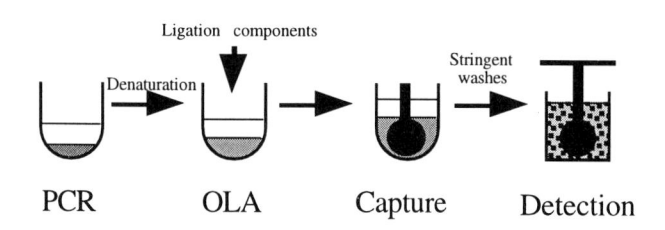

FIGURE 1. Oligonucleotide-ligation analysis of allelic sequence variants in amplified DNA. After target amplification by PCR, a set of three oligonucleotides and a ligase are added, and after a 30 min incubation ligation products are immobilized on a solid support. Ligated molecules are then transferred to a separate well for detection by time-resolved fluorescence measurement.

Protocol

PCR amplification
Ligation reaction
Binding of ligation products to the solid support
Washes
Detection of ligation products

PCR amplification

DNA sequences to be analysed for allelic sequence variants are amplified by PCR, followed by ligase-mediated gene detection. The amplifications can be performed directly in microtiter wells or in Eppendorf tubes, depending on the PCR instruments available. Four µl amplification reactions are used for each ligation reaction. In order to amplify in a total volume of 4µl, mix 2µl of genomic DNA at 2ng/µl in 1x PCR buffer, with 2µl of *Taq* polymerase (0.2U/µl) and primers (2µM each) in 1x PCR buffer. In this manner, fluctuation in buffer composition due to pipetting errors can be avoided. The composition of the 1x PCR buffer is: 50mM Tris-HCl pH8.3, 50mM KCl, and 1.5mM $MgCl_2$.

Ligation reaction

After amplification the reactions are diluted with 6µl water to a volume of 10µl, and the reactions are subjected to a heating step (96°C for 5 min) to denature the PCR products. The temperature is then rapidly lowered to 37°C, and a ligation mix is added to individual amplification reactions.

Ligation mix per reaction:

1) A set of three labelled oligonucleotides (one biotinylated and two allele-specific ones, differentially labelled with europium or terbium chelates; see below) are used at 600fmol each as probes for the ligation reaction.

   0.4mU of T4 DNA ligase
   1.8µl of 50mM Tris-acetate pH 7.5
   50mM magnesium acetate
   250mM potassium acetate
   0.8µl 5M NaCl
   0.2µl 100mM ATP.

2) Ten µl of the mixture is added to each microtiter well after the solutions have reached temperatures below 60°C.

3) Ligation reactions are incubated for 30 min at room temperature or at 37°C.

Binding of ligation products to a solid support

Add 20µl of solution A (1M NaCl, 100mM Tris-HCl pH 7.5, 0.1% Triton X 100) and introduce the streptavidin-coated manifold solid support in each reaction (for a description of the support see Parik et al. in this volume). Incubate on a shaking platform at room temperature for at least 15 min. As an alternative, 2µl per reaction of streptavidin-coated paramagnetic particles (Dynabeads, Dynal AS) can be used as a solid phase.

Washes

Wash the solid support twice with solution A as in the previous step, then wash once with a denaturing solution (0.1 M NaOH, 1M NaCl, 0.1% Triton X100), and finally two washes with solution A.

Detection of ligated products

Add 180µl of europium-fluorescence enhancement solution (0.1M acetate-phthalate pH 3.2, 15mM 2-naphtoyl trifluoroacetone, 50mM tri-N-octylphosphine oxide, and 0.1% Triton X 100; available from Wallac, Finland) to the wells of a microtiter plate and introduce the solid supports. Incubate 10 min on a shaking platform. Remove the supports and record the europium signals in a Delfia Plate Reader Research Fluorometer (Wallac, Finland). Add 20µl of a terbium enhancement solution (100mM 4-(2,4,6-trimethoxyphenyl)-pyridine-2,6-dicarboxylic acid and 1% cetyltrimethylammonium bromide in 1.1M NaHCO$_3$; a gift from Wallac), shake 10 min, and record the terbium signals.

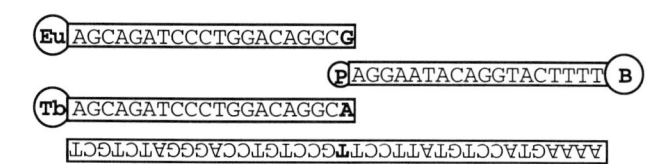

FIGURE 2. Structure of the oligonucleotides used in ligation-mediated analysis of PCR products. The probes shown are designed to identify a mutation in coagulation factor V, the so-called Leiden mutation, associated with deep venous thrombosis. The probe 5'-labelled with europium (Eu) selectively detects the normal allele and that labelled with terbium (Tb) is specific for the mutant sequence. The variant positions in the probes and target sequences is indicated in bold.

Comments

- Oligonucleotides used in this ligation-based assay are complementary to the target sequence and designed so that the analysed mutation lies at the border between two juxtaposed oligonucleotides (Fig. 2). One of the oligonucleotides is complementary to a sequence in common for both the normal and mutated sequence. This oligonucleotide is modified at the 5' end with a phosphate group and at the 3' end with a biotin residue. The other two oligonucleotides, hybridizing immediately upstream of the first one, are identical in sequence except for their 3' ends where one is complementary to a normal and the other to a mutant sequence variant. These two allele-specific oligonucleotides are 5' labelled with detectable groups; that specific for a normal allele carries europium-chelates, and the one specific for the mutant, terbium-chelates.

- The two lanthanide labels selected for this assay, chelates of europium and terbium ions, permit sensitive detection of as little as 0.1µl of amplification reactions and the two colors are well resolved using a commercially available microplate reader. The detectable groups are characterized by an unusual type of fluorescence with a very wide distinction between excitation and emission wavelengths of over 200nm, and extremely long duration of fluorescence after excitation, permitting time-resolved measurement to discriminate against background. The synthesis of the labelled probes is described by Kwiatkowski et al. (1994). Probes can also be modified with chelates by reacting oligonucleotides having reactive amines with a reagent commercially available from Wallac.

- The key component of the fluorescence enhancement solution used for detection of terbium ions is not commercially available. The synthesis is described by Hemmilä (1993). Other, commercially available substances that can be employed are described by Hemmilä (1985). As a further alternative, probes can be labelled with europium and samarium chelates. The fluorescence of both these ions can be recorded in the enhancement solution used for europium, commercially available from Wallac, although samarium ions are detected with approximately 5-fold lower sensitivity than europium ions.

- The method to screen large sets of patient samples for known mutations described here presents several advantages: The ligation reaction accurately distinguishes between sequence variants, and it in fact offers sufficient specificity to detect genes directly in complex genomic DNA samples. The manifold supports used to process reactions greatly simplifies the procedure and reduces the risk for mistakes. Finally, the dual-label design permits internally controlled analysis of each sample, enhancing the precision of the analysis. The method is also useful for scoring biallelic genetic variants in forensics or for genetic linkage analysis. Moreover, by selecting an appropriate control sequence, added to the nucleic acid samples at a known concentration, the presence of specific gene sequences can be quantitated using the same ligase-mediated, dual-color assay format.

References

Hemmilä, I. Time-resolved fluorometric determination of terbium in aqueous solution. *Anal. Chem.* **57**: 1676-1681, 1985.

Hemmilä, I., Mukkala, V.-M., Latvam M., and Kiilholma, P. Di- and tetracarboxylate-derivatives of pyridines, bipyridines and terpyridines as luminogenic reagents for time-resolved fluorometric determination of terbium and dysprosium. *J. Biochem. Biophys. Meth.* **26**: 283-290, 1993.

Kwiatkowski, M., Samiotaki, M., Hurskainen, U., Landegren, U. Solid-phase synthesis of chelate-labelled oligonucleotides: Application in triple-color ligase-mediated gene analysis. *Nucl. Acids Res.* **22**: 2604-2611, 1994.

Parik, J., Kwiatkowski, M., Lagerkvist, A., Samiotaki, M., Lagerström, M., Stewart, J., Glad, G., Mendel-Hartvig, M., and Landegren, U. A manifold support for molecular genetic reactions. *Anal. Biochem.* **211**: 144-150, 1993

Samiotaki, M., Kwiatkowski, M., Parik, J., Landegren, U. Dual-color detection of DNA sequence variants through ligase-mediated analysis. *Genomics* **20**: 238-242, 1994.

LCR
LIGASE CHAIN REACTION

Bruce Wallace*, Luis Ugozzoli, A.A. Reyes, and J. Lowery
*Communicating author

Sickle cell anemia, perhaps the first human disorder to be recognized as a molecular disease, is the result of the mutation of a single A to a T in codon 6 of the β-globin gene. While this mutation causes easily measured changes in the hemoglobin molecule, DNA testing for sickle cell anemia is useful when the β-globin is not expressed in available tissues, when the amount of sample is small, or when one wishes to use a different method for confirmation than was used for screening. Sickle cell anemia clearly represents a paradigm for genetic diseases. Today, gene mutations responsible for a myriad of disorders are being discovered at a rapid rate due, in part, to the human genome project.

DNA diagnostics has been a long standing application of molecular biology. To make the technique more clinically useful, DNA diagnostics must be non-radioactive, rapid, sensitive, specific, and cost effective. One of the major obstacles to overcome has been to develop tests that are both sensitive and non-radioactive. This challenge has been met in a number of ways, including the use of biologically amplified targets (eg rRNA), signal amplification (eg branched DNA), and template amplification (eg PCR and the ligase chain reaction or LCR)(Landegren, 1993).

LCR is a template dependent amplification reaction. The reaction utilizes two pairs of oligonucleotides, one pair complementary to the upper template strand and one pair complementary to the lower template strand. The pairs hybridize to adjacent positions on the template, such that a DNA ligase can join the 5' phosphate of one oligonucleotide to the 3' hydroxyl of the other. In addition to being template dependent, this joining reaction is sequence specific: if the template contains a non-complementary base at the ligation junction, ligation is inhibited. This property makes LCR allele specific. The joined product of one round of ligation can serve as a template in subsequent cycles, thus resulting in an exponential accumulation of products.

Unlike PCR, LCR does not synthesize new DNA during the process. Instead, LCR results in the conversion of the shorter pairs of oligonucleotides into longer products by ligation. The products of ligation can be detected in a non-radioactive detection reaction. Here we describe a protocol for the detection of the sickle cell mutation using LCR (Wu and Wallace, 1989)

Protocol

Isolation of genomic DNA

LCR amplification

Non-radioactive detection of LCR products

Interpretation of results

Isolation of genomic DNA

Genomic DNA is extracted from dried blood spots on Guthrie cards. Three mm punches from the spots are placed into 1.5ml tubes, resuspended in 1.25ml of lysis buffer (0.32M sucrose, 10mM Tris-HCl pH 7.4, 5mM $MgCl_2$, 1% Triton X-100), and shaken at 4°C for 15 min. The tubes are then microcentrifuged for 15 sec and the supernatants are carefully eliminated. This lysis step is repeated once. Next, 100µl of Instagene (Bio-Rad Laboratories, Hercules, CA) is added to each sample. The tubes are boiled for 8 min and mixed on a vortex mixer for 10 sec. After 15 sec microcentrifugation, the supernatants are carefully collected and used for the LCR reactions.

LCR amplification

LCR assays are designed for the gene target of interest. By way of example, the following oligonucleotides were synthesized and used in LCR for analysis of the sickle cell mutation:

MD074, 5'- ACATGGTGCACCTGACTCCTG

MD073, 5'- AGGAGAAGTCTGCCGTTACTTTTGGCACTGG CCGTCGTTTTAC

MD076, 5'- TGGAGAAGTCTGCCGTTACTTTTGGCACTGG CCGTCGTTTTAC

MD075, 5'- CAGGAGTCAGGTGCACCATGTT-Biotin

MD077, 5'- CAGTAACGGCAGACTTCTCCT

MD078, 5'- CAGTAACGGCAGACTTCTCCA

1) LCR assays are carried out by ligation of oligonucleotide primers specific for the normal β-globin gene (β^A)(MD074, MD075, MD077, and MD073) and primers specific for the sickle cell allele (β^S) (MD074, MD075, MD078, and MD076) in the presence of genomic DNA template extracted from the filters. Primers MD075, MD073, and MD076 are phosphorylated using ATP and T4 polynucleotide kinase.

2) Reactions are performed in a 25µl volume containing 20mM Tris-HCl pH 7.6, 10mM MgCl$_2$, 100 mM KCl, 0.1% Triton X-100, 100fmol of each LCR primer, 1mM NAD, 6.6U of *Taq* ligase (New England BioLabs), and 12.5µl of genomic DNA template.

3) To check for the presence of template-independent ligation, two negative controls are prepared. In the first control, genomic DNA isolated from an Epstein-Barr virus transformed cell line containing a homozygous deletion of the β-globin gene (βth/βth) is used as a template, and in the second control, salmon sperm DNA (250ng) is the DNA template.

4) The reactions are subjected to 25 cycles of heating and cooling using the following program: 94°C for 2 min and 62°C for 4 min for one cycle and 94°C for 30 sec and 62°C for 4 min for the next 24 cycles.

Non-radioactive detection of LCR products

LCR products can be detected in a microtitre plate format by a non radioactive technique. In this procedure, one of the two LCR primers that are complementary to the sense DNA strand of the β-globin gene (MD075), is biotinylated at the 3' end and phosphorylated at the 5' end, while one of two primers complementary to the anti-sense strand (MD073 or MD076) is phosphorylated at the 5' end and contains a DNA tail at the 3' end which is complementary to an oligonucleotide-alkaline phosphatase conjugate (UP-AP secondary probe, Bio-Rad Laboratories, Hercules).

The LCR product is a double-stranded DNA molecule with biotin on one strand and a DNA tail, which can be detected with the UP-AP conjugate, on the other. The alkaline phosphatase probe acts on a substrate nicotinamide adenine dinucleotide phosphate whose product initiates a secondary cyclic enzyme reaction, which amplifies the initial product producing a final colored product, formazan.

1) An aliquot (1-10µl) of the LCR is diluted in 1x SSC (0.15 M NaCl, 0.015M sodium citrate) in a streptavidin-coated microwell (Bio-Rad Radias B12 plate, Part No. 430 1006) to a final volume of 50µl.

2) The plate is incubated at 37°C for 2 hr. The wells are washed five times with 300µl of Radias Well Wash (Bio-Rad Part No. 439 3001).

3) The UP-AP conjugate (50 μl of a 1nM solution in 3M NaCl, 1x SSC) is added to each well and incubated at 37°C for 1 hr.

4) Each well is washed five times with 300μl Radias Well Wash.

5) Fifty μl of alkaline phosphatase substrate (Bio-Rad Part No. 439 1003) is added and the wells incubated at 37°C for 1 hr.

6) Fifty μl of amplifier reagent (Self, 1985)(Bio-Rad Part No. 439 1001) is added, and the plate immediately transferred to a plate reader set to read at 490nm. The rate of color development is determined by doing a kinetic read at 20 sec intervals for 5 min.

Interpretation of results

The template- and allele-dependence of LCR is exemplified in the following table, where X1 and X2 are alleles of gene X and S1 and S2 are LCR oligonucleotide sets, specific for X1 and X2, respectively. Because of the template and allele specificity of LCR, and the fact that the products of LCR can be directly measured, LCR is ideally suited for the diagnosis of hemoglobinopathies and other diseases caused by known genetic defects. The template- and allele-specificity of the LCR concept is summarized as follows:

Template	LCR with S1	LCR with S2
X1/X1	+	−
X2/X2	−	+
X1/X2	+	+
None	−	−

References

Landegren, U. Molecular mechanics of nucleic acid sequence amplification. *Trends Genet.* **9**: 199-204, 1993.

Self, C.H. Enzyme amplification - a general method applied to provide an immunoassisted assay for placental alkaline phosphatase. *J. Imm. Meth.* **76**: 389-393, 1985.

Wu, D.Y. and Wallace, R.B. The ligation amplification reaction (LAR): Amplification of specific DNA sequences using sequential rounds of template-dependent ligation. *Genomics* **4**: 560-569, 1989.

CHAPTER • 21

UHG
HETERODUPLEX AND UNIVERSAL HETERODUPLEX GENERATOR ANALYSIS

N.A.P. Wood* and J.L. Bidwell

*Communicating author

The altered gel migration properties of heteroduplex DNA molecules, formed by hybridization of mismatched DNA strands, can be used to monitor DNA sequence differences. The technique can also be used to study known alleles of genes, such as those of the human major histocompatibility complex, by coamplifying a DNA sample from a patient with a universal heteroduplex generator (UHG) DNA molecule. The sequence of these UHGs is selected so that native gel electrophoresis of heteroduplexes generated in the amplification process will produce well-resolved bands due to mismatched positions in the heteroduplex molecules (Fig. 1).

Protocol

Isolation of DNA from peripheral blood leukocytes

Isolation of DNA from dried blood spots

PCR

Nondenaturing polyacrylamide minigel electrophoresis

Construction of universal heteroduplex generators (UHGs)

Analysis of known point mutations using UHGs

Procedure for crossmatching with UHG

Isolation of DNA from peripheral blood leukocytes

1) Centrifuge 10ml whole blood for 10 min at 1300x g.
2) Transfer the buffy layer (approximately 1ml) to a 10ml disposable plastic conical tube. Add 8ml red cell lysis buffer (0.144M ammo-

nium chloride, 1mM sodium hydrogen carbonate). Mix and let stand for 20 min at ambient temperature.

3) Centrifuge for 10 min at 1300x g. Remove the red cell lysate as near to the white cell pellet as possible. Resuspend the white cell pellet in 3ml nuclei lysate buffer (10mM Tris-HCl pH 8.2, 0.4M NaCl, 2mM EDTA pH 8.0). Add 0.6ml 1x proteinase K solution (2mg/ml proteinase K in 1% w/v SDS, 2mM EDTA pH 8.0, 5x proteinase K solution contains 10mg/ml proteinase K in 1% w/v SDS, 2 mM EDTA pH 8.0), and 0.2ml 10% w/v sodium dodecyl sulphate (SDS). Mix well.

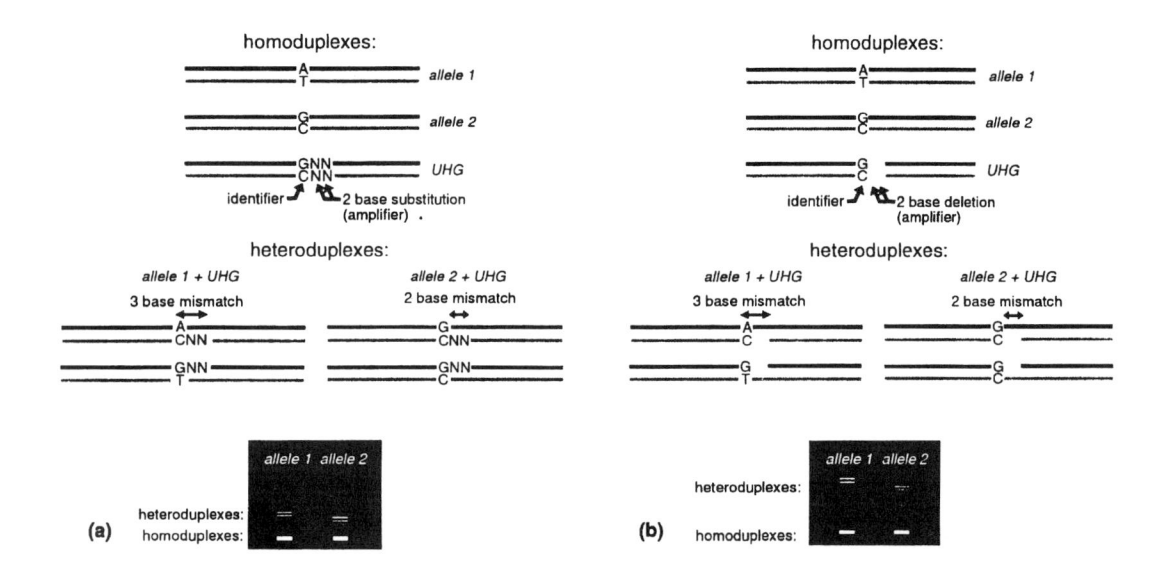

FIGURE 1. Principle of UHG analysis. Two alleles (1 and 2) of a hypothetical gene are shown which differ by a single base pair (A-T to G-C mutation) but which are otherwise matched along the length of the region amplified by PCR. Coding strands are shhown in black, noncoding strands in grey. In (a) a UHG is constructed which has the same sequence as allele 2 except that two nucleotides contiguous with the polymorphism (the amplifiers) are changed in sequence to mismatch with both alleles. In the heteroduplex, this has the effect of creating a 3-base mismatch between allele 1 and the UHG, and a 2-base mismatch between allele 2 and the UHG. The higher number of mismatches with allele 1 has the effect of increasing gel retardation over that seen with allele 2 (see lower schematic diagram over banding patterns). In (b) a UHG is constructed as in (a) except that a 2-base deletion is used in stead of a 2-base substitution, as the amplifier. In the heteroduplexes, this also has the effect of creating a 3-base mismatch between allele 1 and the UHG, and a 2-base mismatch between allele 2 and the UHG. Here, however, the mobilities of all heteroduplexes (see lower schematic diagram over banding patterns) are more highly retarded than in (a), due to the net loss of two anionic phosphate groups, and conformational distorsions (see text). In practise, a UHG would contain multiple identifiers and amplifiers, so that more than two alleles could be simultaneously identified.

4) Incubate for 18 hr in a 37°C waterbath (Note: 5x proteinase K solution may be substituted in place of 1x: in this case incubate for 3 hr at 55°C).

5) Add 1ml 6M NaCl, shake vigorously for 15 sec.

6) Centrifuge for 25 min at 1300-1500x g.

7) Pipette the supernatant into a clean disposable plastic conical tube, avoiding the pellet. Add 8ml absolute ethanol. Cap the tube and mix gently by inversion.

8) Remove the precipitated DNA with a sealed glass Pasteur pipette, squeeze out the excess ethanol, and redissolve the DNA in 0.1-0.3ml double distilled water. Store at 4°C until required.

9) For DNA assay, add 5µl of the DNA solution to 495µl of double distilled water and measure the optical density at 260nm. The concentration of DNA (in µg/µl) is 5x the OD value.

Isolation of DNA from dried blood spots

DNA can also be rapidly isolated from dried blood spots, normally used in neonatal screening tests (Guthrie spots).

1) Excise a 2mm^2 area of a blood spot from the paper and place in a 0.5ml microcentrifuge tube.

2) Add 25µl of a 1:10 dilution of AmpliTaq™ reaction buffer and heat at 96°C for 15 min.

3) Add a further 25µl of a 1:10 dilution of AmpliTaq reaction buffer, mix, and centrifuge at 13,000x g for 15 min.

4) Use 30µl of the supernatant in the PCR.

PCR

With many PCR-based applications, inter-laboratory and inter-instrument variation in results is sometimes considerable. For this reason, the methods below should be used as a baseline. Optimal volumes and PCR parameters may need to be determined empirically.

The methods below are recommended for the Perkin-Elmer 480 and other first-generation thermal cyclers: for the Perkin-Elmer System 9600, total reaction volumes can be proportionately reduced.

1) Set up the PCR in a 0.5ml microcentrifuge tube, as follows:

    Genomic DNA 500ng   5.0µl*
    5' PCR primer (5µM stock)   10.0µl
    3' PCR primer (5µM stock)   10.0µl
    AmpliTaq reaction buffer   10.0µl

dNTP mix (10mM each dNTP)   2.0µl
Double-distilled water (ddH$_2$O)   62.0µl

*If DNA is obtained from dried blood spots as above, use 30µl of supernatant and reduce the ddH$_2$O volume to 37.0µl.

2) Vortex the tube, then centrifuge briefly. Overlay with 50µl paraffin oil.

3) To hot-start the reaction, transfer to a dry-block of thermal cycler set at 72°C and leave for 10 min.

4) Add the Ampli*Taq* DNA polymerase (2U: 0.4µl from a 1:10 dilution of Ampli*Taq* reaction buffer) by passing the pipette tip through the mineral oil.

5) Subject samples to PCR using 35 cycles of:

Denaturation: 1 min at 94°C
Primer annealing: 2 min at annealing temp
Primer extension: 2 min at 72°C.

For the final cycle, increase the 72°C primer extension time to 10 min. With the Perkin-Elmer instruments, use a THERMO-CYCLE file to achieve suggested ramping times. Alternatively, faster times and a variable number of cycles may be used. In this case, to maximise heteroduplex formation in the last PCR cycle, heat to 94°C for 5 min, then allow to cool slowly over 10 min to the annealing temperature before the final primer extension step.

Nondenaturing polyacrylamide minigel electrophoresis

The following electrophoresis conditions are suitable for resolving heteroduplexes formed during PCR. Details of the electrophoresis conditions for UHG analysis are found under "Procedure for crossmatching with UHG".

Heteroduplexes are resolved in nondenaturing (native) polyacrylamide gels. For speed and convenience, this is performed using minigel electrophoresis. The conditions described below are for the Bio-Rad Mini-Protean II gel electrophoresis system.

1) Pour a 12% nondenaturing polyacrylamide (29:1 acrylamide:bisacrylamide) minigel, using 1x Tris-borate, EDTA (TBE) running buffer.

2) Add 8-10µl of PCR product to 2µl 6x sucrose gel loading buffer, containing bromophenol blue and xylene cyanole FF.

3) Separate by electrophoresis at 200 volts for 90-95 min. The exact time will need to be determined experimentally: the homoduplex

(leading) band should be allowed to migrate to within 0.5cm of the end of the gel.

4) Stain the gel for 20 min in 1x TBE containing 0.5µl/mg ethidium bromide and examine using a 302nm UV transilluminator.

Troubleshooting note: Reducing genomic DNA and/or Ampli*Taq* concentration in the PCR can reduce background fluorescence or nonspecific reactions if this is a problem.

Construction of universal heteroduplex generators (UHGs)

Synthesis of long oligonucleotides

Oligonucleotides are synthesized using standard phosphoramidite chemistry on high coupling-efficiency support columns. (ABI 40 nanomole polystyrene or Millipore 0.25mole membrane supports).

Construction of UHGs from a single long oligonucleotide (80 to 130bp)

UHG suitable for crossmatching analysis can be generated from the amplification of 10ng of a crude long oligonucleotide in a standard amplification reaction.

Construction of UHG from two or more long oligonucleotides (140 to 250bp)

1) In a standard amplification reaction of 100µl, amplify 2.5µl of each oligonucleotide at 2.5 µM (Wood et al., 1993). The choice of DNA polymerase can be critical: *Taq* polymerase is suitable for most applications but if the UHG sequence is known to contain regions difficult to synthesise such as those having greater than 3 G residues in a row, then a DNA polymerase with a proof reading activity should be used.

2) Check amplification on an agarose minigel.

3) Pool the products from 4 to 6 reactions and resolve on 8-12% acrylamide minigels (use wide slot purification gels and adjust the gel composition according to the size of the UHG).

4) Visualize UHG band by ethidium bromide staining, and excise the band.

5) Elute DNA from gel fragments in an electroelution device (AE-6580 max yield NP Atto, Japan).

6 Precipitate DNA using standard ethanol/ammonium acetate procedure, wash the pellet in 70% ethanol and dry.

7) Redissolve the pellet of UHG DNA in 100-150µl of sterile water.

8) Dilute an aliquot of UHG serially (6 dilutions 10^{-1}, 10^{-2}, 10^{-3}, 10^{-4}, 10^{-5} and 10^{-6}).

9) Optimize the re-amplification of diluted UHG (normal range between 10^{-4} and 10^{-6}).

Analysis of known point mutations using UHGs

UHGs designed to allow identification of separate alleles are amplified separately from patient samples. As an example an UHG used for analysis of the human HLA DP locus is shown in Fig. 2.

PCR from genomic DNA is carried out essentially as above. For amplification of the UHG, either M13mp18 clones or purified DNA are used as a source. Thermal cycling parameters are as above, and the mixture is:

UHG: M13mp18 DNA (1/100 dilution from supplied stock) or purified UHG (1/1000 dilution from supplied stock)	1.0µl
5' primer (5µM stock)	10.0µl
3' primer (5µM stock)	10.0µl
AmpliTaq reaction buffer	10.0µl
dNTP mix (10mM each dNTP)	2.0µl
ddH$_2$O	66.5µl

Hot-start the PCR with 1.0µl of diluted AmpliTaq and continue as above.

Procedure for crossmatching with UHG

1) Amplify genomic test samples, control samples, and sufficient UHG samples for analysis (one 100µl reaction provides enough material for 6 crossmatches).

2) Check amplification fidelity and strength on agarose minigel.

3) Aliquots (normally 10-15µl) of each amplicon, genomic and UHG are denatured together at 94°C for 3 min then allowed to slowly to re-anneal and cross hybridize by ramping to 37°C over 30 min.

4) Homoduplexes and heteroduplexes are resolved on 6cm. by 8 cm Mini Protean gel systems (Bio-Rad)(Bidwell and Hui, 1990).

   Gel composition, non-denaturing polyacrylamide gels at 2.6% bis acrylamide cross-linking;

80bp	20%	total acrylamide	
120bp	17.5%	"	"
160bp	15%	"	"
200bp	12%	"	"

5) Banding patterns and analysis; gels are post electrophoretically stained with either ethidium bromide or SYBR green (Molecular Probes), UV illuminated and photographed using Polaroid 667 film.

UNIVERSAL HETERODUPLEX GENERATOR

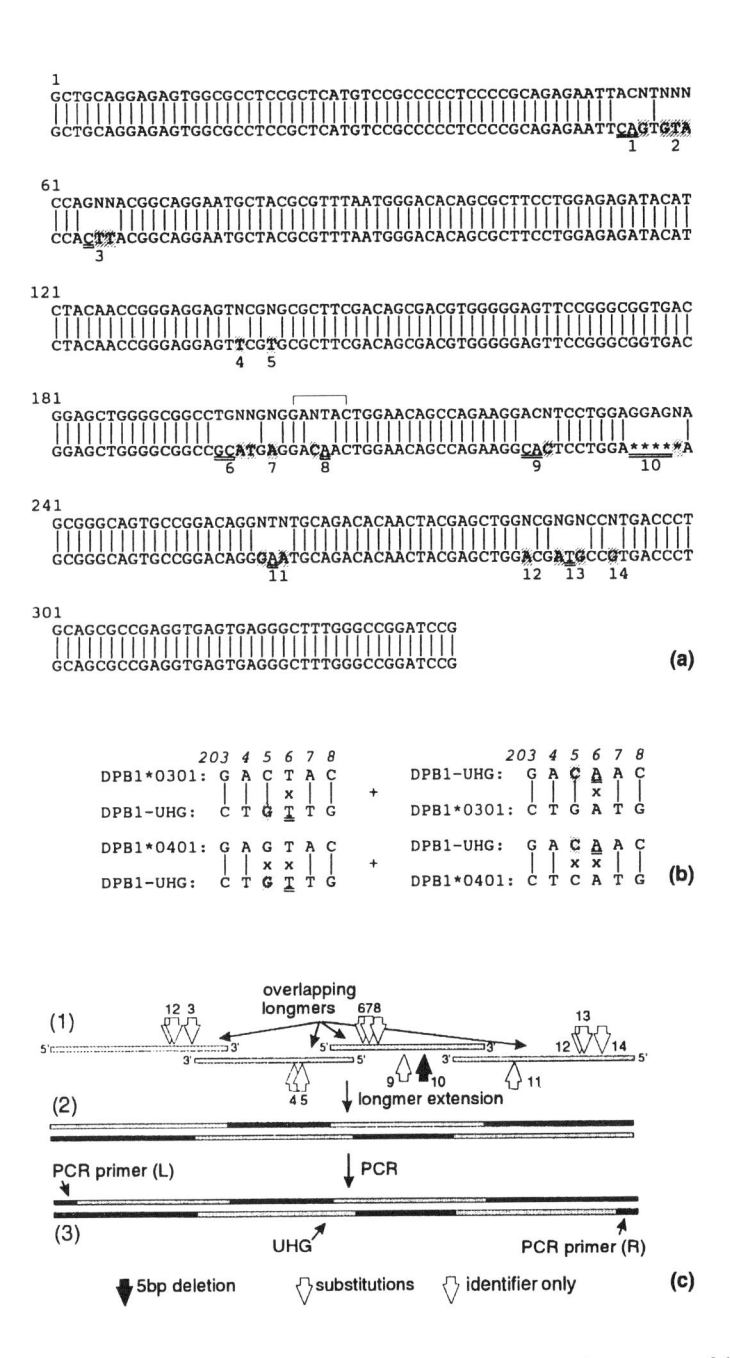

```
1
GCTGCAGGAGAGTGGCGCCTCCGCTCATGTCCGCCCCCTCCCCGCAGAGAATTACNTNNN
||||||||||||||||||||||||||||||||||||||||||||||||||||||
GCTGCAGGAGAGTGGCGCCTCCGCTCATGTCCGCCCCCTCCCCGCAGAGAATTCAGTGTA
                                                    1   2

61
CCAGNNACGGCAGGAATGCTACGCGTTTAATGGGACACAGCGCTTCCTGGAGAGATACAT
|||   ||||||||||||||||||||||||||||||||||||||||||||||||||||
CCACTTACGGCAGGAATGCTACGCGTTTAATGGGACACAGCGCTTCCTGGAGAGATACAT
    3

121
CTACAACCGGGAGGAGTNCGNGCGCTTCGACAGCGACGTGGGGGAGTTCCGGGCGGTGAC
||||||||||||||||| | |  |||||||||||||||||||||||||||||||||||
CTACAACCGGGAGGAGTTCGTGCGCTTCGACAGCGACGTGGGGGAGTTCCGGGCGGTGAC
                 4 5

181                          ┌──┐
GGAGCTGGGGCGGCCTGNNGNGGANTACTGGAACAGCCAGAAGGACNTCCTGGAGGAGNA
|||||||||||||||||  | || |||||||||||||||||||| |||||||||
GGAGCTGGGGCGGCCGCATGAGGACAACTGGAACAGCCAGAAGGCACTCCTGGA*****A
              6   7   8                        9         10

241
GCGGGCAGTGCCGGACAGGNTNTGCAGACACAACTACGAGCTGGNCGNGCCNTGACCCT
|||||||||||||||||||| | ||||||||||||||||||||| | || || |||||||
GCGGGCAGTGCCGGACAGGAATGCAGACACAACTACGAGCTGGACGATGCCGTGACCCT
                   11                         12 13  14

301
GCAGCGCCGAGGTGAGTGAGGGCTTTGGGCCGGATCCG
||||||||||||||||||||||||||||||||||||||
GCAGCGCCGAGGTGAGTGAGGGCTTTGGGCCGGATCCG                    (a)


        203 4 5 6 7 8              203 4 5 6 7 8
DPB1*0301: G A C T A C      DPB1-UHG:  G A C A A C
           | |   x | |                 | |   x | |
DPB1-UHG:  C T G T T G      DPB1*0301: C T G A T G

DPB1*0401: G A G T A C      DPB1-UHG:  G A C A A C
           | | x x | |                 | |   x x | |
DPB1-UHG:  C T G T T G      DPB1*0401: C T C A T G      (b)
```

FIGURE 2. Design of an UHG for analysis of HLA DPB1. (a) The structure of the relevant gene segment with the variable positions are shown. (b) The base pairing pattern of two alleles in a small segment of the gene, bracketed in (a), with the corresponding segment of an UHG. (c) Strategy for construction and cloning of the UHG from several oligonucleotides.

References

Bidwell, J.L. Hui, K.M. Human HLA-DR/Dw allotype matching by analysis of HLA-DRB gene PCR product polymorphism (PCR "fingerprints"). *Technique* **2**: 93-100, 1990.

Wood, N. Tyfield, L. Bidwell, J. Rapid classification of phenylketonuria genotypes by analysis of heteroduplexes generated by PCR-amplifiable synthetic DNA. *Hum. Mutation* **2**: 131-137, 1989.

DNA SEQUENCE ANALYSIS

SOLID-PHASE
DNA SEQUENCING

Mathias Uhlén and Joakim Lundeberg*
*Communicating author

DNA sequencing strategies used in large genome projects or in routine diagnostic analyses require increases in efficiency to reduce cost and avoid manual steps. Immobilization of the DNA to be sequenced on solid supports can facilitate automation of the sequencing steps. Therefore, protocols for solid-phase sequencing have been developed and are now in routine use. Our solid-phase protocol uses paramagnetic beads with covalently coupled streptavidin, Dynabeads M-280 (Dynal AS, Norway), for the capture of biotin labelled DNA fragments (Hultman et al., 1991).

Incorporation of biotin can be performed in several ways, however, this is preferably done using one biotinylated PCR primer. A schematic outline of the procedure is shown in Fig. 1, where the target DNA for amplification can be either cloned or genomic DNA. The amplified and biotin-labelled products are allowed to bind to the monodisperse beads, coupled with streptavidin, and immobilization takes place in a few seconds. Thereafter, nucleotides, primers, the polymerase etc are washed away by applying a magnetic field to collect the beads with the bound DNA to the side of the tube, and aspirating the liquid to waste. Alkali treatment of the bead/DNA-complex separates the DNA strands, and two single-stranded sequencing templates are obtained. One remains immobilized onto the beads, and the other is recovered in the supernatant and neutralized.

These manipulations eliminate the need for any precipitations, extractions, or filtration steps, thus facilitating rapid preparation of large numbers of pure sequencing templates. Subsequently, the two single strands are used as templates in standard Sanger reactions, preferably using T7 DNA polymerase or Sequenase. Both the first part of the procedure, the DNA-capture and single strand preparation, as well as the subsequent primer annealing and sequencing reaction steps, are suitable for automation.

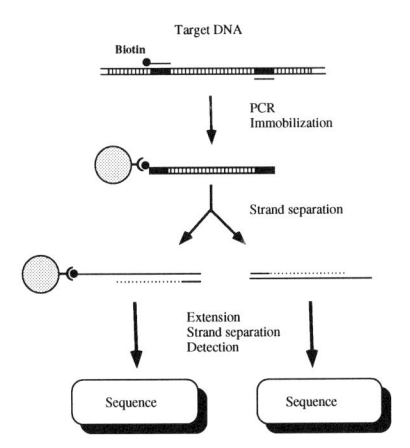

FIGURE 1. Schematic outline of the solid phase sequencing concept.

The robotic handling of immobilization and sequencing reactions in a microtiter format has been performed in our lab using two Biomek-1000 robotic workstations (Beckman Instruments, USA), equipped with a neodymium-iron-boron permanent magnet (Dynal AS, Norway) and a HCB-1000 heating/cooling device (Beckman Instr.), respectively. An ABI 800 Catalyst (Applied Biosystems, USA), fitted with an experimental magnetic station, has been used together with the 373A DNA Sequencer from Applied Biosystems.

With no further manipulations the samples are ready to be loaded on the electrophoresis gel. This can be done automatically, by transferring the microtiter plate to a pipette robot (Biorad, USA), for direct loading on the electrophoresis gel. Raw data is generated by laser-mediated fluorescent on-line detection of the dideoxy fragments by using automated sequencing instruments from Pharmacia Biotech (A.L.F., Uppsala, Sweden) or from Applied Biosystems. Both instruments utilize computerized registration and processing of raw data. These data can subsequently be sent to a SUN graphic workstation which allows the use of powerful software for further sequence analysis, i.e. Staden, PHYLIP etc.

Protocol

PCR

Preparing the beads

Immobilization of the PCR product

Strand separation

Single strand recovery

Manual solid phase DNA sequencing using the A.L.F. sequencer

PCR

1) Pick a colony with a pipette and suspend it in 10µl lysis buffer (1x PCR buffer), incubate in a thermocycler at 99°C for 5 min.
2) Spin down the cell debris.
3) For a 50µl PCR, mix 5µl 10x PCR buffer, 5µl dNTP mix, and 2µl of the appropriate primer solution (5 pmol of each PCR primer), provided in the kit. Adjust the volume to 49µl with dH$_2$O.
4) Take 1µl of the colony lysate or 1-2fmol of purified DNA to the PCR mixture and add 0.2µl (1U) Taq polymerase.
5) Cover the PCR mixture with 20µl mineral oil. The primers provided in the Starter Kit from Dynal are designed for use at high annealing temperatures. We recommend using a two step PCR, with annealing/extension at 72°C, resulting in more specific amplification products.

Preparing the beads

1) Twenty µl of washed Dynabeads M-280 Streptavidin (10mg/ml) are used for each sequencing reaction (the beads may be washed in bulk and thereafter aliquoted).
2) The beads are washed once with 1 vol of binding solution. Washing is done by placing the tube in the magnet and removing the buffer that the beads are provided in.
3) Add binding solution, turn the tube around 4-5 times in the magnet, and allow the beads to sediment at the side of the tube.
4) Remove the supernatant with a pipette.
5) Resuspend the beads in twice their original volume of binding solution, 40µl per PCR sample (a final concentration of 5mg/ml).

Immobilization of the PCR product

1) Transfer the amplification reaction to a 40µl aliquot of the washed paramagnetic beads.
2) Incubate at room temperature for 15 min. The beads should be resuspended 2-3 times during the incubation.

Strand separation

1) Collect the beads using a magnet and remove the supernatant with a pipette.
2) Wash the beads once with 50µl of binding solution and once with 50µl of 1x TE buffer. The beads and the immobilized double stranded DNA can be stored at 4°C for several weeks.

3) Remove the TE buffer avoiding droplets at the walls and bottom of the tube and resuspend the beads in exactly 10µl 0.1M NaOH solution.
4) Incubate at room temperature for 5 min.

Single strand recovery

Collect the beads to the side of the tube and transfer exactly 10µl of the NaOH supernatant (non-biotinylated strand) to a fresh tube. Wash the beads once with 50µl 0.1M NaOH, once with 50µl binding solution, and once with 50µl TE buffer. Remove the 1x TE, without leaving any droplets, and adjust the volume with water to the volume appropriate for your sequencing protocol. Neutralize the NaOH supernatant (containing the non-biotinylated strand) with 3µl of 0.33M HCl, mix immediately with a pipette and adjust the volume with water according to your sequencing protocol.

Manual solid phase DNA sequencing using the A.L.F. sequencer

1) Adjust the volume to 13µl with dH$_2$O.
2) Add 2µl (1 pmol) fluorescent primer.
3) Add 2µl annealing buffer.

    Mix gently with a pipette and heat at 65°C for 10 min..

    Mix gently again and leave to cool at room temperature for at least 10 min, mix gently with a pipette occasionally during the cooling.

4) Add 1µl of extension buffer and mix gently.
5) Dilute the T7 DNA polymerase to 1.5U/µl using cold enzyme dilution buffer; 2µl of this diluted stock will be required for each template to be sequenced. Mix by gentle pipetting and keep on ice. Do not remove the T7 polymerase from -20°C for more than a few seconds.
6) For each template label four Eppendorf tubes or, if many samples are to be sequenced, microtiter plate wells: "A", "C", "G", and "T". Dispense 2.5µl of the corresponding sequencing mix into the wells.
7) Warm the dispensed sequencing mixes at 37°C for at least 1 min.
8) Add 2µl of diluted T7 DNA polymerase to the annealing mix (from step 4) and mix gently with a pipette. Immediately add 4.5µl of this mixture to each of the warmed sequencing mixes.
9) Incubate the reactions for 5 min at 37°C.
10) Add 5µl of stop solution to each reaction and mix gently.
11) Just prior to loading the gel, heat the reactions at 95°C for 2 min, cool and load them on the gel.

Materials

Solutions and chemicals

10x PCR Buffer (100mM Tris-HCl pH 8.3 (at 20°C), 20mM MgCl$_2$, 500mM KCl, 1% Tween 20)

Ampli *Taq* (Perkin Elmer, Sweden)

Magnetic beads (Dynabeads M-280 Streptavidin (Dynal AS, Norway), aliquoted into 200μl batches)

Binding solution (10mM Tris-HCl pH 7.5, 1mM EDTA, 2M NaCl)

TE buffer (10mM Tris-HCl pH 7.5, 1mM EDTA)

0.1M NaOH (freshly made)

0.33M HCl (freshly made)

Annealing buffer (280mM Tris-HCl pH 7.5, 100mM MgCl$_2$), sequencing primer (diluted with water to 0.14 OD/ml (0.5pmoles/μl))

Nucleotide mixes, 840μM of each dNTP (c^{7}dGTP is used instead of dGTP) and 3.65μm of the specific ddNTP

50 mM NaCl and 40 mM Tris-HCl pH 7.5

Extension Buffer (300 mM citric acid pH 7.0, 318 mM dithiothreitol, 40 mM MnCl$_2$)

Stop solution (shake 5g Amberlite MB-1 in 100ml formamide and 300mg dextran blue for 30 min. Filter through 0.45μm membrane).

Equipment

Magnet MPC-E (Dynal AS, Norway)

Thermocycler PE 9600 (Cetus, USA)

37°C water bath

45°C and 65°C incubator.

References

Hultman, T., Ståhl, S., Hornes, E., Uhlén, M. Direct solid phase sequencing of genomic and plasmid DNA using magnetic beads as solid support. *Nucl. Acids Res.* **17**: 4937-3946, 1989.

Hultman, T., Bergh, S., Moks, T., Uhlén, M. Bidirectional solid-phase sequencing of in vitro-amplified plasmid DNA. *BioTechniques* **10**: 84-93, 1991.

CHAPTER • 23

MANIFOLD SEQUENCING

Arild Lagerkvist and Ulf Landegren*

*Communicating author

DNA sequence analysis is a fundamental technique in mutation detection. It permits distinction between causative mutations and polymorphisms that can be found in the same gene. For DNA fragments where multiple sequence variants are expected, such as MHC genes and variable microbial sequences, it is the method of choice, and can provide information on the relatedness between individuals or organisms from which the samples were derived. However, the numerous processing steps required, and the difficulty of detecting heterozygous mutations, have limited its use in diagnostics. We describe here a comb-shaped support

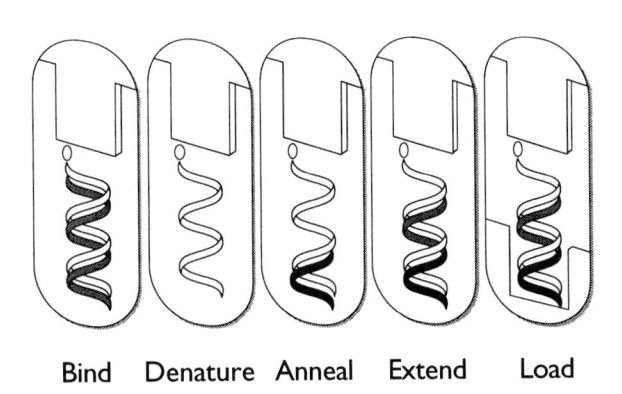

Bind   Denature   Anneal   Extend   Load

FIGURE 1. Manifold sequencing. Biotinylated amplified DNA fragments are bound to a set of supports, shaped like teeth of gel combs, allowing up to 256 individual reactions to be handled in parallel. After removing the complementary strands by denaturation, a sequencing primer is hybridized to bound templates, and extension reactions are performed, followed by direct loading of the products using the supports.

that can be used to capture and then transfer sets of biotinylated PCR products between reaction wells for denaturation, annealing of the sequencing primer, extension reaction, and gel loading on a fluorescence DNA sequencer (Fig. 1). In this manner, up to 256 samples can be processed in parallel (Lagerkvist et al., 1994), providing excellent quality sequence results, and also avoiding the risk of sample mix-up and contamination.

The binding capacity of the support is expanded by grafting Sepharose particle with conjugated streptavidin on to the plastic surface of the manifold device (Parik et al., 1993), see separate protocol in this volume by Parik et al.

Protocol

PCR amplification
Binding of PCR products
Sequencing reaction
Gel loading

PCR amplification

Perform the PCR with the biotinylated primer at 0.1μM. The non-biotinylated end of the PCR product should include the sequence of the oligonucleotide to be used as a sequencing primer. This can be incorporated as a 5'-extension of the non-biotinylated primer. As an alternative, the sequencing primer can be designed to hybridize internal to the non-biotinylated amplification primer. In the latter case the sequencing reaction affords some added specificity in that nonspecific PCR products amplified with the first primer pair will not serve as templates in the sequencing reaction.

Eighty μl of the PCR product is used for binding to four comb teeth for one sequencing reaction.

Binding of PCR products

Supports can be made as described (Parik et al., 1993) or may be obtained from Pharmacia Biotech (Autoload Sequencing Combs). PCR products of up to 800bp can be bound to the comb-shaped support in sufficient amounts to be sequenced. The binding capacity of the support decreases with the length of the PCR product. For shorter fragments 2 hr binding is adequate. Fragments above 400bp should be allowed at least 5 hr to bind. Optimal binding is achieved by over-night incubation at 42°C.

MANIFOLD SEQUENCING

Binding can be done at room temperature but the efficiency is somewhat lower. The presence of 1M NaCl enhances binding.

1) Aliquot 20µl of 5M NaCl in long reaction wells that fit four comb teeth.
2) Add 80µl of PCR products to each well.
3) Insert the combs.
4) Seal the plates with the combs in a plastic bag with a damp paper towel inside.
5) Incubate at 42°C over night.

The combs with bound PCR products can be stored at +4°C if not used directly.

Sequencing reaction

Denaturation

1) Aliquot 0.15M NaOH, 0.15M NaCl in long wells or in a beaker. Immerse the combs for at least 5 min.
2) Rinse the combs once by immersing the tips of the teeth in a beaker with 10mM Tris-HCl pH 7.5, 50mM NaCl, and shake gently for 30 sec.

Annealing

1) For each template, mix 1.5pmol of a fluorescein-labelled sequencing primer with 100µl 10mM Tris-HCl pH 7.5, 50mM NaCl and add in a long well.
2) Insert the combs and incubate at 65°C for 10 min, then continue the incubation at room temperature for at least a few more minutes.

Extension

1) Make up four reaction mixtures in Eppendorf tubes. Put the tubes on ice.

   ddH$_2$O   8µl
   d/ddNTP mix   8µl
   Buffer mix   4µl

   The buffer mix is made up of Buffer 1, Buffer 2, and Buffer 3 (see below), mixed in equal volumes. The mixture of the three buffers should not be stored.

2) Add 4U of T7 DNA polymerase per reaction and distribute 20µl of the extension mixes in four wells.
3) Heat the reaction plate at 42°C and insert the combs.

4) Incubate for 10-30 min.

Stop

Put the reaction plates with the combs on ice until loading.

The reactions can be stored for longer time at -20°C. The combs should be kept in the reaction wells.

Gel loading

1) Preheat the gel to 50°C.
2) Fill the wells of the gel with deionized formamide.
3) Insert the combs and incubate for 10 min.
4) Pull out the combs gently.
5) Set the gel temperature at 45°C and start the run.

Reagents

Buffers	Reagent	Add	Stock
Buffer mix 1	Tris-HCl pH 7.5	311µl	1M
	ddH$_2$O	689µl	
Buffer mix 2	Dithiothreitol	177µl	1M
	ddH$_2$O	823µl	
Buffer mix 3	MnCl$_2$	62µl	1M
	Isocitrate	460µl	1M
	ddH$_2$O	478µl	

- Sterile filter the buffers. Store at -20°C.

Termination mixes

Solutions have to be set up for each of the four ddNTPs, the asterisc (*) means that only one of the ddNTPs is present in the respective mixes A, C, G, and T. If band compressions are a problem, dGTP can be replaced by c7dGTP at the same concentration.

Reagent	Final conc.	Add	Stock
dATP	1mM	10µl	100mM
dCTP	1mM	10µl	100mM
dGTP	1mM	10µl	100mM
dTTP	1mM	10µl	100mM
ddNTP*	5µM	1µl	5mM
NaCl	50mM	10µl	5M

Tris-HCl	40mM	40µl	1M
ddH$_2$O		909µl	

Comments

The technique presented here simplifies the problem of generating and processing large sets of high-quality sequencing templates. The same devices are also useful for radioactive sequencing, and they should also prove of use in techniques for mutation scanning by mismatch analysis.

References

Lagerkvist, A., Stewart, J., Lagerström, M., Parik, J., and Landegren, U. Manifold sequencing: rapid processing of large sets of sequencing reactions. *Proc. Natl. Acad. Sci. USA* **91**: 2245-2249, 1994.

Parik, J., Kwiatkowski, M., Lagerkvist, A., Samiotaki, M., Lagerström, M., Stewart, J., Glad, G., Mendel-Hartvig, M., and Landegren, U. A manifold support for molecular genetic reactions. *Anal. Biochem.* **211**: 144-150, 1993.

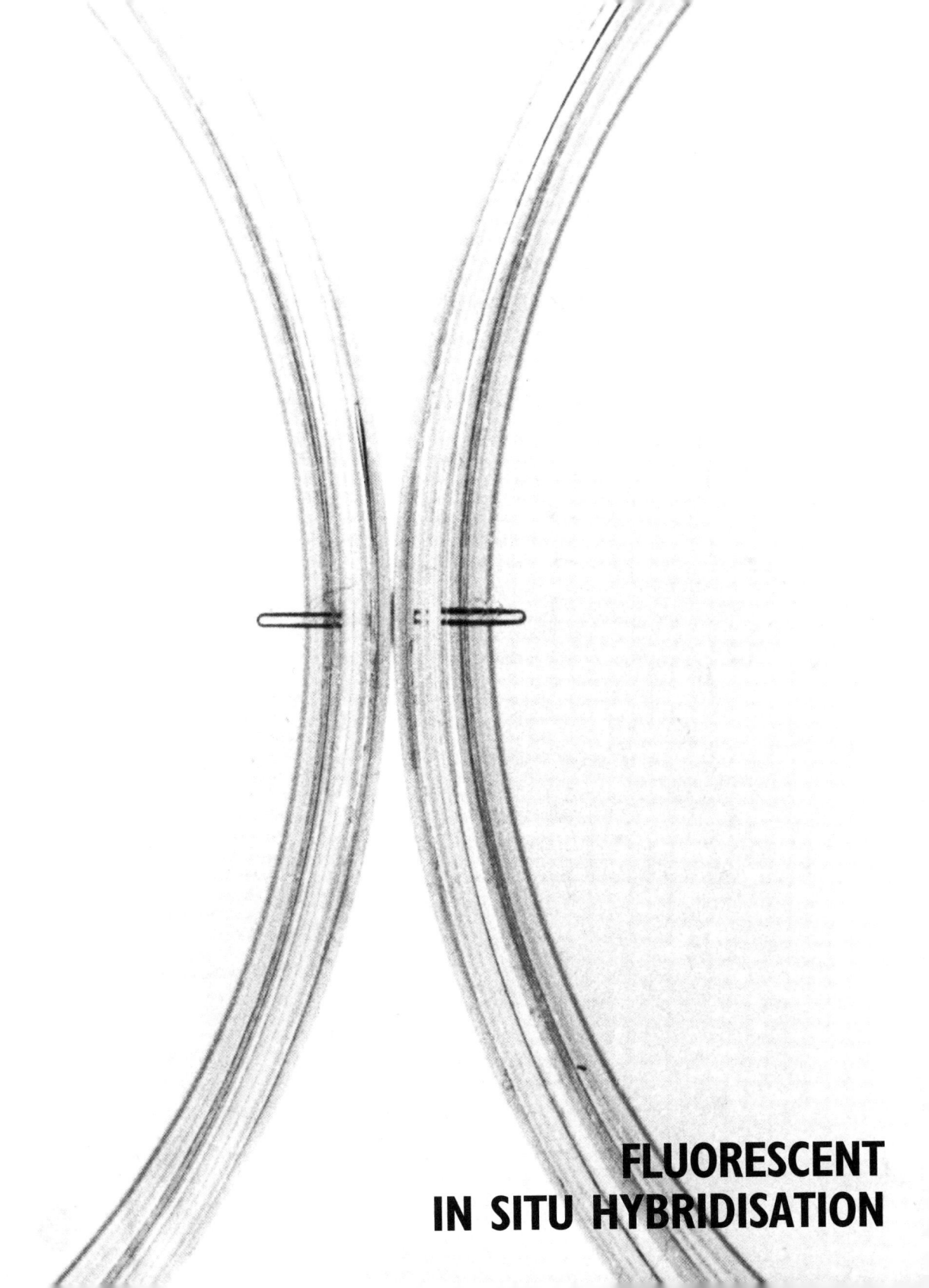

FLUORESCENT IN SITU HYBRIDISATION

SENSITIVE FISH
USING BIOTIN TYRAMIDE-BASED DETECTION

Ton Raap*, Gert-Jan Van Ommen and Joop Wiegant
*Communicating author

Biotin-tyramide is used to amplify signals in fluorescent in situ hybridization (FISH) analysis. After in situ hybridization with a digoxigenin- or biotin-labelled probe, the slides are incubated with peroxidase-conjugated anti-digoxigenin or streptavidin. Biotin-tyramide is then applied as a peroxidase substrate to generate and locally deposit many biotin molecules. Finally, fluorochrome-labelled avidin is used to visualize the biotin deposition (Raap et al., 1995).

Protocol

Nick-translation and preparation of probe solutions
Preparation of metaphase chromosomes
In situ hybridization to chromosomes
Biotin-tyramide based detection

Nick-translation and preparation of probe solutions

Prepare the following labelling mixture in an Eppendorf tube kept on ice:

Filtered distilled water	26µl
Nick-translation buffer (10x)	5µl
Dithiothreitol (DTT; 0.1M)	5µl
Nucleotide mix (0.5mM dA, dG, dC; 0.1mM dT)	4µl
Digoxigenin-11-dUTP (1mM)	2µl
Probe DNA (1µg)	1µl
DNA polymerase I (10U/µl)	2µl
DNase I (1:1000 from a 1mg/ml stock)	5µl

Note: For labelling with biotin- or fluorochrome-labelled nucleotides replace the digoxigenin-11-dUTP by 2μl of a 1mM solution of the relevant modified dUTP.

Incubate for 2 hr at 15°C and stop the reaction by alcohol precipitation as described below.

Prepare the hybridization solution, including Cot-1 competitor DNA to suppress hybridization to repeats:

1) Add 50μg herring sperm DNA, 50μg yeast RNA, and a 50-fold excess of human Cot-1 DNA.
2) Add 0.1 volume of 3M sodium acetate (pH 5.5) and mix.
3) Add 2.5 volume of 100% ethanol (-20°C), mix well and place the solution for 30 min on ice.
4) Spin at 12000x g for 30 min at 4°C.
5) Remove supernatant and resuspend the pellet in 10μl of 50% deionized formamide/2x SSC/50mM sodium phosphate/10% dextran sulphate (pre-warmed to 37°C).
6) Dissolve the DNA for 15 min at 37°C.
7) Denature the DNA by placing the sample(s) for 5 min at 75°C.
8) Place the DNA for 1 min on ice, spin down.
9) Pre-anneal by placing the probe solution for 30 min at 37°C.

Preparation of metaphase chromosomes

1) Add 10ml of heparin blood to 100ml of chromosome culture medium (Boehringer) and divide over 4 Falcon culture flasks.
2) After 72 hr of culture, add 80μl 0.0025% colcemid per flask.
3) Incubate at 37°C for 2 hr.
4) Pool the contents of the 4 flasks and mix gently.
5) Divide the contents over 15ml tubes (10ml of chromosome culture per tube).
6) Centrifuge for 8 min at 700rpm.
7) Aspirate the supernatant till 1ml is left.
8) Resuspend the pellet and add 10ml hypotonic buffer while vortexing carefully.
9) Mix by shaking gently and place the tubes for 20 min in a 37°C waterbath.
10) Centrifuge for 8 min at 700rpm.
11) Aspirate the supernatant to 0.5ml.

12) Resuspend the pellet.

13) Add fixative in the following way (fixative is 3 parts methanol + 1 part acetic acid, prepared freshly):

I) Fill a Pasteur pipette with fixative.

II) Add a little bit of fixative to the cell suspension and immediately aspirate it back into the Pasteur pipette.

III) Add some more fixative to the cell suspension and aspirate it back into the Pasteur pipette.

IV) Repeat a couple of times until the whole content of the pipette has been added to the cell suspension. In this manner the cells are fixed while being constantly kept in movement. Repeat these manipulations until 10ml of fixative has been added per tube. After the first Pasteur pipette of fixative has been added to the cell suspension the actions can be done more rapidly. The first addition of fixative is the most critical one.

14) Let the tubes stand for 10 min.

15) Centrifuge 7 min at 800rpm, aspirate the supernatant, leaving 0.5ml, and resuspend the pellet.

16) Repeat the fixation 3 or 4 times. Let the tubes stand for 20 to 30 min between each fixation step and store at -20°C.

17) Prepare metaphase spreads by dropping. Check the quality of the chromosome preparations by phase contrast microscopy. If cytoplasm is present, rinsing the slides with 70% acetic acid shortly before it dries, may help to remove it.

18) Store metaphase preparations in 70% ethanol at 4°C.

In situ hybridisation to chromosomes

Pretreatments

1) Apply 120µl RNase A (100µg/ml in 2x SSC) on the slide and cover it with a 24x60mm glass coverslip.

2) Incubate for 1 hr at 37°C in a moist chamber (1000ml beaker containing paper tissues moistened with water).

3) Wash the slide 3 x 5 min in 2x SSC at room temperature (RT).

4) Wash for 5 min in 2x SSC at 37°C.

5) Place for 10 min in a 5mg pepsin/100ml 0.01M HCl solution at 37°C.

6) Wash for 5 min in 1x PBS at RT.

7) Place the slide for 10 min in a 1% formaldehyde/1x PBS/50mM $MgCl_2$ solution at RT.

8) Wash for 5 min in for PBS at RT.
9) Dehydrate the slide through an ethanol series (70-90-100% ethanol, 5 min each).
10) Air dry.

Hybridization

1) Apply 120µl of 70% deionized formamide/2x SSC/50mM sodium phosphate pH 7 on the slide.
2) Cover the solution with a 24x60 mm coverslip.
3) Denature the chromosomal DNA 2-3 min on a 80°C hot plate and remove the coverslip.
4) Place the slide directly in ice-cold 2x SSC and wash for 2 min.
5) Dehydrate in 70% ethanol (-20°C) 2 x 5 min.
6) Dehydrate further through an ethanol series (90 and 100% ethanol, RT, each for 5 min).
7) Place the slide on a plate at 37°C, and leave it there to dry.
8) Apply 10µl of the pre-annealed probe mixture per target area on the pre-denatured slide.
9) Cover the probe mixture with a 18x18mm coverslip.
10) Hybridize overnight at 37°C in a moist chamber (1000ml beaker containing paper tissues moistened with 50% formamide/2x SSC pH 7).

Post hybridization washes

1) Prepare a 50% formamide/2x SSC pH 7 solution and warm to 45°C.
2) Prepare a 0.1x SSC solution and warm to 60°C.
3) Place the slide in the formamide solution and remove the coverslip by gently shaking.
4) Wash 3x5 min in the formamide solution at 45°C.
5) Wash 3x5 min in 0.1x SSC at 60°C.

Biotin tyramide-based detection

1) Wash the slide for 5 min in 0.1M Tris-HCl, 0.15M NaCl, 0.05% Tween 20 pH 7.4 (TNT).
2) Block with 0.1M Tris-HCl, 0.15M NaCl, 0.5% Boehringer Blocking Reagent pH 7.4 (TNB) for 30 min at 37°C.
3) Wash briefly with TNT.
4) Incubate peroxidase-conjugated sheep-anti-digoxigenin (Boehringer)

diluted 1:200 (in case of biotin probes, streptavidin-peroxidase (Vector) 1:500) in TNB for 30 min at 37°C.

5) Wash 3 x 5 min with TNT.

6) Incubate the biotin-tyramide substrate solution for e.g. 10 min at RT. The stock biotin-tyramide solution of 1mg/ml in dimethylsulfoxide (Nen-DuPont) is diluted 500- to 1000-fold in 0.2M Tris-HCl, 10mM imidazole pH 8.8, and fresh H_2O_2 added to a final concentration of 0.01%.

7) Wash with TNT.

8) Incubate with streptavidin-FITC (or -Texas Red or -AMCA; all from Vector at working dilutions of 1:500) in TNB for 30 min at 37°C.

9) Wash with TNT.

10) Dehydrate as above in an ethanol series, air dry, and mount in antifading agent (Vectashield from Vector), containing diamidino phenyl indole (DAPI; 40ng/ml) or propidium iodide (PI; 100ng/ml).

9) Inspect the result under a fluorescence microscope equipped with the appropriate filter sets.

Reference

Raap, A. K., van de Corput, M.P.C., Vervenne, E.A.W., van Gijlswijk, R.P.M., Tanke, H.J., Wiegant, J. Ultrasensitive FISH using peroxidase-mediated deposition of biotin- or fluorochrome tyramides. *Hum. Mol. Genet.* 4: 529-534, 1995.

CHAPTER 25

FIBER-FISH

Mervi Heiskanen and Aarno Palotie*

*Communicating author

This high-resolution fluorescent in situ hybridization (FISH) protocol describes the preparation of, and in situ hybridization to extended DNA fibers from melted pulsed-field gel electrophoresis (PFGE) sample blocks. FISH on DNA fibers allows ordering of probes at a resolution of 1-300kb and also provides information on the physical distances. In addition to mapping of the normal genome, the method is also suitable for the analysis of genomic rearrangements, such as inversions and amplifications (Fig. 1).

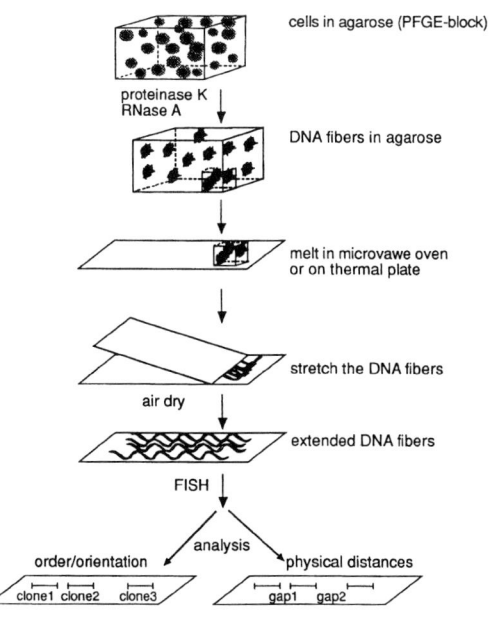

FIGURE 1. Preparation of extended DNA fibers for analysis by fiber FISH. Cells in agarose blocks are digested, the blocks are melted, and fibers of DNA are extended along the slide using another slide. After drying the fibers are subjected to analysis by FISH.

Protocol

Coating of slides with poly-L-lysine

Preparation of target DNA

Labelling of probes and
preparation of hybridization mixture

Hybridization and post-hybridization washings

Hybridization detection

Coating of slides with poly-L-lysine

The slides are coated with poly-L-lysine to improve DNA attachment to the glass surface.

Dip slides in racks for 30 sec in each of:

1) 0.2M HCl,
2) distilled water,
3) acetone.
4) Air dry at room temperature (RT).
5) Dip in 0.15% gelatin/0.03% sodium azide for 5 min.
6) Air dry overnight at RT.
7) Dip the slides in 0.2% poly-L-lysine solution for 10 min, rinse in distilled water for 30 sec, and air dry 1 hr at RT.
8) Repeat step 7.
9) Air dry overnight at RT.
10) Store the slides at 4°C.

Preparation of target DNA

PFGE blocks are prepared following protocols described for PFGE.

1) Wash cells twice with 1xPBS (lymphocytes separated by density gradient centrifugation or cultured cell lines) and resuspend in 1xPBS at a final concentration of $2x10^6$ cells/100µl.
2) Add an equal volume of 1.9% low melting point agarose (FMC BioProducts, Rockland, ME, USA) in PBS, dispense the cells into block formers, and allow to set.
3) Incubate the blocks in a solution of 2mg/ml proteinase K, 50mM EDTA, 1% N-lauroylsarcosine at 50°C for 2-4 days. The proteinase digestion is followed by several washings with 1x TE pH 7.5.

4) Place a small piece of the block (1/8 of a 100µl block) on a microscope slide.
5) Add 20µl of water to the block.
6) Melt the block in a microwave oven, 30 sec at 700W.
7) After melting, DNA is extended on a slide using another microscope slide (Fig. 1). Do not fix the slides.
8) Denature the slides in 70% formamide/2x SSC for 4 min at 74°C.
9) Dehydrate the slides in an ice cold ethanol series 70%, 90% and 100%, 3 min each.
10) Air dry.

Labelling of probes and preparation of hybridization mixture

Nucleotide mix: 0.2mM each of dATP, dGTP, dCTP in 500mM Tris-HCl pH 7.5, 50mM $MgCl_2$, 100mM 2-mercaptoethanol.

Nick translation reaction (50µl)

Nucleotide mix	5.0µl
bio-11-dUTP (0.4mM) or	2.5µl
dig-11-dUTP (1mM)	0.5µl
Bovine serum albumin (BSA; 10mg/ml)	0.5µl
Probe DNA (1µg)	xµl
DNA polymerase I/DNase I mix	
(0.4U/µl DNA pol I, 40pg/µl DNase I)	5.0µl
Filtered distilled water	ad 50µl

1) Incubate 1.5 hr at 15°C, then inactivate the enzyme by incubating for 15 min at 80°C.
2) Check the probe fragment size by gel electrophoresis. The ideal fragment size is 300-1000bp.
3) Hybridization mixture for one slide: 7.5µl of nick translated probe (150ng), 5µg herring sperm DNA, 2µg human Cot-1 DNA. Add 2.5x volume of 100% ice cold ethanol, mix and keep the solution at -20°C for 30 min.
4) Centrifuge for 30 min at 4°C and wash the pellet with 70% ethanol (-20°C).
5) Resuspend the pellet in 30µl of 50% deionized formamide/10% dextran sulphate/2x SSC.
6) Dissolve the DNA by incubating for 20 min at 37°C.
7) Denature the probe by boiling for 5 min.
8) Suppress repetitive sequences by preannealing for 15 min at 37°C.

Hybridization and post-hybridization washings

Add 30μl of probe solution to a slide, cover with 24x50mm coverslip, and seal the coverslip with rubber cement. Hybridize in a moist chamber at 37°C overnight.

Wash the slides in 50% formamide/2x SSC 3 x 5 min, 2x SSC 2 x 5 min, 0.5x SSC 1 x 5 min and briefly in 4x SSC/0.05% Tween-20, all at 45°C.

Detection

1) Block with 5% BSA/4x SSC Tween-20 for 15 min at 37°C.
2) Biotinylated probes are detected with avidin-TRITC (Vector, 1:1000) and the signal is amplified by incubation with biotinylated anti-avidin (Vector, 1:200) and another layer of avidin-TRITC.
3) Digoxigenin-labelled probes are detected using mouse anti-digoxigenin (Boehringer Mannheim, 1:300), sheep anti-mouse Ig-FITC (Sigma, 1:200), and donkey anti-sheep Ig-FITC (Sigma, 1:200).
4) Incubations are made in a moist chamber for 20 min at 37°C. Between incubations the slides are washed in 4x SSC Tween-20, 3 x 5 min at 45°C.
5) Slides are stained with DAPI (5μg/ml) and evaluated under a fluorescence microscope.

Comments

The most critical step in the protocol is the melting of the agarose block. The melting time and the amount of water added should be optimized for different owens. If the melting is not efficient, unmelted agarose left on the slide causes background problems.

References

Heiskanen, M., Karhu, R., Hellsten, E., Peltonen, L., Kallioniemi, O.P., and Palotie, A. High resolution mapping using fluorescence in situ hybridization to extended DNA fibers prepared from agarose-embedded cells. *Biotechniques* 17: 928-933, 1994.

Heiskanen, M., Hellsten, E., Kallioniemi, O-P., Mäkelä, T., Alitalo, K., Peltonen, L. and Palotie, A. Visual mapping by fiber-FISH. *Genomics*, **30**: 31-36, 1995.

Laan, M., Kallioniemi, O.-P., Hellsten, E., Alitalo, K., Peltonen, L., and Palotie, A. Mechanically stretched chromosomes as targets for high-resolution FISH mapping. *Genome Res.* 5: 13-20, 1995.

PADLOCK PROBES
FOR *IN SITU* DETECTION

Mats Nilsson* and Ulf Landegren

*Communicating author

A novel probe design, the padlock probe, may provide the necessary specificity and stability to identify the location of specific gene sequence variants in blots and microscopic specimens. The same probe design should also prove useful in library screening, both on membranes and in solution.

A padlock probe is a probe molecule, designed to include two target-complementary segments, connected by a linker segment (Fig. 1). The target-complementary segments can hybridize immediately next to each other. When a DNA ligase is added, the hybridized ends of the probe become ligated, converting the linear probe into a circle (Nilsson et al., 1994). Since the hybridized probe is wound around the target DNA molecule, circularized probes are linked to the target molecule like links in a chain. This design offers the exquisite target sequence-specificity of

FIGURE 1. Structure of a padlock probe. A linear probe molecule with detectable groups (stars) is designed to include two target-complementary segments at either end. Upon hybridization to the target sequence, these two ends are brought in juxtaposition and may be joined by ligation, converting the probe to a circular molecule that is catenated to the target sequence.

a probe ligation reaction, combined with the possibility to wash bound probes under superstringent, denaturing conditions. This will allow allele distinction in situ with very low nonspecific signals.

Contents

Probe construction
Hybridization and ligation
Detection

Probe construction

The probes used for detection are designed to have two target-complementary segments, usually approximately 20 bases long each, and separated by a spacer of around 50 bases. As a general rule spacers slightly longer than the sum of the two probe segments have been used to bridge the hybridizing segments. The spacer is generally made up of T-residues. Alternatively, hexaethylene glycol (HEG) residues (Jäschke et al., 1993) are incorporated during synthesis. In the spacer, residues suitable for detection via e.g. biotin-streptavidin interactions can also be included.

The 5' end of the probe oligonucleotide must be phosphorylated in order to be ligatable. This phosphate can be introduced during chemical synthesis, or it can be added to the probe as a radiolabelled phosphate, using polynucleotide kinase, for radioactive detection. It is important that most or all probes are of full length, since truncated molecules will occupy target sites without being able to ligate and contribute to the signal after denaturing washes. We recommend isolation of full-length products by gel purification from denaturing polyacrylamide gels.

Hybridization and ligation

DNA molecules in samples to be analyzed using padlock probes should preferably be circularily closed or alternatively long and/or bound at multiple points to a solid support (e.g. a glass slide or a membrane) to ensure that the cyclized probes do not slip off the target during denaturing washes. Before the hybridization the DNA sample is denatured using alkali, heat, or formamide.

Hybridization is generally performed with the probe at 2-50fmol/µl for 30 min to 16 hr in 300mM NaCl, 30mM Na citrate (2x SSC), 5x Denhardt's solution (1mg/ml each of Ficoll, polyvinylpyrrolidone, and bovine serum albumin), and salmon sperm DNA at 0.5mg/ml at 37°C.

After a brief wash in ligation buffer, ligation is performed at room temperature or at 37°C for 1 hr in 10mM Tris-HCl pH 7.5, 10mM Mg(Ac)$_2$,

50mM KAc, 1mM ATP, and 0.15U/µl T4 DNA ligase. Thermostable ligases can also be used.

Washes are performed using 2x SSC with different formamide concentrations at elevated temperatures or 0.1 M NaOH at room temperature. Unligated probes can also be removed using single strand-specific exonucleases such as exonuclease VII (5'- and 3'- specific) or spleen phosphodiesterase (5'- specific). If radioactive phosphates are used as label, alkaline phosphatase efficiently removes the signal from unligated probe molecules.

Detection

The signal can be detected by recording radioactivity through autoradiography or by generating chemiluminescence with an avidin-peroxidase conjugate that binds to biotinylated probes. Alternatively, fluorescence can be visualized from fluorophores directly incorporated in the probe or conjugated to avidin, secondarily bound to biotinylated probes.

Chemiluminescent detection can be used to localize a biotinylated probe, directed to target DNA immobilized on a nylon membrane (PALL). After ligation and wash, the membranes are washed once more in 2x SSC, 2% SDS for 30 min, and then incubated in 0.05µg/ml of a streptavidin-horseradish peroxidase conjugate (Boehringer-Mannheim) in 2x SSC, 2% SDS. The membranes are rinsed in phosphate-buffered saline for 30 min and then soaked in ECL solution (Amersham) for 1 min. The chemoluminescent signal is recorded on X-omat-S film (Kodak).

Fluorescent signals can be recorded in metaphase chromosome in situ hybridization and ligation reactions, using biotinylated probes. After the circularization reaction and washes, the slides are rinsed in PN buffer (0.1 M NaH_2PO_4, 0.1% Nonidet P-40, adjusted to pH 8.0 with 0.1 M Na_2HPO_4). Bound probes are visualized using fluorescein-labelled avidin, followed by a layer of biotinylated anti-avidin antibodies, both at 5µg/ml (Vector Laboratories), and a second layer of fluorescein-labelled avidin. The sensitivity of this analysis is adequate for target sequences repeated in tandem, but insufficient for single copy targets. All incubations are performed in PN buffer with 5% non-fat milk at 37°C for 20 min followed by three washes in PN buffer at room temperature, 5 min each. The metaphase chromosomes are counterstained with propidium iodide and photographed using a Nikon microscope.

Comments

Circularizable probes require that two probe segments hybridize in juxtaposition on a target sequence, and, once hybridized a ligase can link the probe to the target sequence, permitting superstringent washes to remove background. With suitably sensitive detection of probes, even single copy target segments of just 40 bases should be detectable and allelic sequences may be distinguished. The technique should find applications for determining the localization of specific sequences along chromosomes or chromatin fibers, and RNA molecules in tissue segments could be analyzed. Using the same probe construction, specific recombinant molecules may be captured from complex populations for !library screening.

References

Jäschke, A., Fürste J.P., Cech, D., Erdmann, V.A. Automated incorporation of polyethylene glycol into synthetic oligonucleotides. *Tetrahedron Lett.* **34**: 301-304, 1993.

Nilsson, M., Malmgren, H., Samiotaki, M., Kwiatkowski, M., Chowdhary, B.P., and Landegren, U. Padlock probes: Circularizing oligonucleotides for localized DNA detection. *Science* **265**: 2085-2088, 1994.

PROTEIN LEVEL ASSAYS

PTT
PROTEIN TRUNCATION TEST

Rob B. Van Der Luijt*, Frans B.L. Hogervorst, Johan T. Den Dunnen, P. Meera Khan, and Gert-Jan B. Van Ommen

*Communicating author

The protein truncation test (PTT) rapidly detects mutations that interrupt the reading frames of genes. The technique is based on in vitro-coupled transcription and translation of T7 promoter-modified PCR- or RT-PCR-amplified coding sequences. The principle of PTT is summarized in the figure. In brief, templates for PTT are generated by PCR, using cDNA synthesized by reverse transcription of mRNA ((RT)-PCR). An alternative approach, feasible for large exons with a continuous reading frame (such as *APC* exon 15 or *BRCA1* exon 11) is to use genomic DNA directly for the PCR. During PCR a 36-base pair extension, encoding the bacteriophage T7 promoter sequence as well as an eukaryotic translation initiation signal, is added to that end of the PCR-product which corresponds to the amino terminus. Simultaneous transcription and translation of amplification products is performed in a rabbit reticulocyte lysate system using radiolabelled amino acids. Translation products are subsequently resolved by SDS-PAGE and detected by autoradiography.

Protocol

Isolation of RNA from blood or cultured cells

Synthesis of reverse transcribed DNA

PCR amplification of cDNA or genomic DNA

In vitro transcription and translation

SDS-PAGE analysis

Detection of translation products

Primers for *APC* and *BRCA1* mutation detection

PROTEIN-LEVEL ASSAYS

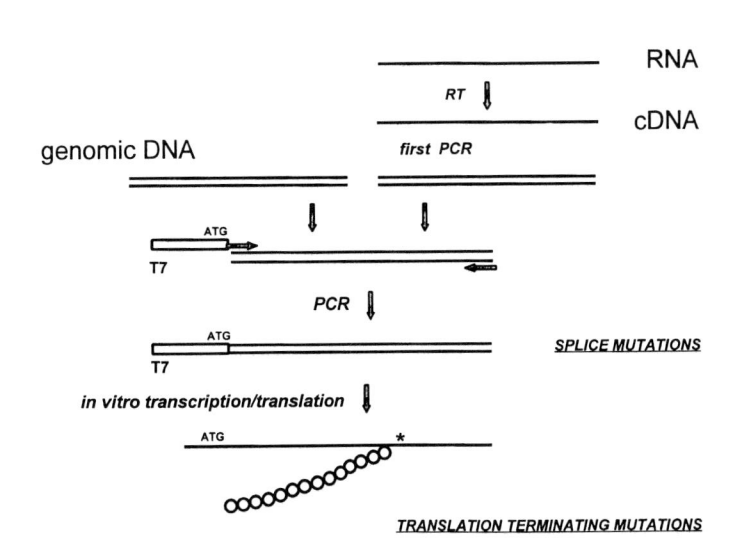

FIGURE 1. Protein truncation test. Genomic or reverse-transcribed mRNA sequences are amplified, incorporating a 36-bp extension upstream of the amplified gene segment. This sequence allows in vitro transcription and translation of the segment, followed by gel analysis of its protein product.

Isolation of RNA from blood or cultured cells

Peripheral blood lymphocytes

1) Take one 10ml tube of blood. We have good results using EDTA-tubes.
2) Fill two 15ml tubes with 5ml Histopaque 1077 (Sigma, St. Louis, MO).
3) Add slowly 5ml of blood on top of the Histopaque.
4) Spin at 2000rpm for 20 min at room temperature (swing out rotor, no brakes)

- After centrifugation 4 layers are visible:

  The top layer contains serum and thrombocytes;
  the second layer is white and contains lymphocytes;
  the third layer is colourless and is Histopaque (sometimes red because of lysed erythrocytes);
  the fourth layer contains erythrocytes.

5) Remove most of the serum layer and collect the second layer using a plastic Pasteur pipette.

6) Wash the lymphocytes with 10ml of cold and sterile phosphate buffered saline (PBS).
7) Spin at 1500rpm for 10 min. A small white, sometimes red coloured pellet is visible.
8) Remove supernatant by inverting the tube.
9) Resuspend the lymphocytes by tapping the tube, and put on ice.

Tissue culture cells

1) Take a tissue culture flask (75cm^2) with cells grown to 80-90% confluency.
2) Pipette off the culture medium.
3) Wash the flask carefully with sterile and cold PBS.
4) Add 5ml of PBS, put the flask on ice and isolate the cells by scraping with a sterile wiper.
5) Cells can also be collected by trypsinization.
6) Put the cells in a tube, add 5ml to the flask and continue scraping.
7) Collect the cells, centrifuge at 1500rpm for 10 min.
8) Remove supernatant and put the cells on ice.

When nonadherent cells are used, take an aliquot (5-10x10^6), spin the tube, and resuspend them in PBS and spin again.

Isolation of total RNA

For rapid and simple isolation of RNA we use RNAzol B (Cinna/Biotecx Labs Inc. Houston, TX). Minimal handling of the sample is required and the procedure takes 1-1.5 hr.

1) Continue directly after isolating the cells with the isolation of total RNA.
2) Briefly resuspend the sample by flicking the tube.
3) Add RNAzol B and lyse the cells by passing the lysate a few times through the pipette tip. For 5ml of blood or one 75cm^2 tissue culture flask we use 1.3ml of RNAzol B.
4) Transfer the lysate to a 1.5-2ml Eppendorf tube.
5) Add 0.1ml chloroform per 1ml of homogenate, shake vigorously for 15 sec, and put on ice for 5 min.
6) Centrifuge the suspension at 12,000x g for 15 min at 4°C. Two phases are formed: an upper colourless aqueous phase, and a lower blue phenol/chloroform phase.

7) Transfer the aqueous phase (0.6-0.7ml) to a fresh tube and add an equal volume of isopropanol. Store the samples on ice for 15 min. At this moment it is possible to store the samples at 4°C until further use.

8) Centrifuge the samples for 15 min at 12,000x g, 4°C. The RNA precipitate is visible as a pellet at the bottom of the tube.

9) Carefully remove the supernatant and wash the pellet with 200µl 70% ethanol.

10) Centrifuge 10 min at 12,000x g and carefully remove all of the supernatant.

11) Air dry the pellet, not for too long, otherwise the pellet will not dissolve anymore.

12) Dissolve the pellet in 50-100µl of TE (10mM Tris-HCl pH 7.5, 0.1mM EDTA).

13) Analyze an aliquot (5-10%) in an RNA gel (1.5% agarose in TBE).

14) Briefly treat the electrophoresis tank (gel apparatus, tray and comb) with 1M NaOH for 1 hour. Wash extensively with sterile water.

15) Pour a TBE gel using sterile TBE and water. Don't forget ethidium bromide.

16) Add to the RNA aliquot an equal amount of 2x formamide gel loading buffer.

17) Incubate for 5 min at 65°C and load on the gel.

18) Inspect the gel on an UV light box after electrophoresis to evaluate the amount and quality of the isolated RNA.

19) Precipitate the RNA by adding 10% vol/vol 3M NaAc pH 5.3 and 2.8 vol ethanol.

20) Store at -20°C or -70°C.

- Always use sterile pipettes, tubes and solutions (if possible).
- Keep the samples cold and wear gloves.

Synthesis of reverse transcribed DNA

To generate first strand cDNA we use the Superscript kit of Life Technologies (Gaithersburg, MD).

1) Take 1-3µg RNA in ethanol in sterile Eppendorf tube, spin 10 min 12,000rpm at 4°C.

2) Remove the ethanol with a sterile pipette and air dry the pellet (not for too long).

3) Add 2µl random primer (0.5µg/µl, Promega, Madison, WI) and 30µl TE, mix and incubate for 10 min at 65°C, put directly on ice.

4) Add: 12µl 5X RT-buffer (Life Technologies)
   6µl 0.1M dithiothreitol (DTT) (Life Technologies)
   6µl 10mM dNTP's (Pharmacia, Uppsala, Sweden)
   1µl 40U/µl RNasin (Promega)
   2-3µl 200U/µl Superscript, MMLV reverse transcriptase (Life Technologies).

5) Mix and incubate 60 min at 42°C.

6) Store on ice for further use or put at -20°C.

PCR amplification of cDNA or genomic DNA

Perform the PCR with a primer pair, of which the 5' end of the sense primer is extended by the following 36-base sequence: 5'-nnTAATACGA-CTCACTAT-AGGAACAGACCACCATGG-3'. Ensure that the correct reading frame is maintained when selecting the gene-specific sequence of the sense primer. For some experiments the addition of restriction sites at the 5'- end of the primer might be useful. PCR-fragments ranging in size between 1 and 2kb are usually convenient to analyse by PTT.

RT-PCR

Many genes are not normally expressed in peripheral blood lymphocytes, however, some illegitimate transcription takes place. We observed that the level of ectopic transcription varies among individuals. Because blood samples from patients and carriers can be obtained without much difficulty, it is also often the only source of material available. In most cases it is necessary to use nested PCR in order to obtain a sufficient amount of the RT-PCR product.

First round PCR

1) Make a mix containing the following per sample:

   2.5µl 10x PCR buffer
   1µl forward primer (20pmol/µl), 1µl reverse primer (20pmol/µl)
   1.5µl dNTPs (25mM)
   2.5µl DMSO (100%)
   0.5µl BSA (10mg/ml, Pharmacia)
   0.2µl Ampli*Taq* (5U/µl, Roche, Branchburg, NJ)
   10.8µl H$_2$O

2) Add 5 µl RT product and 2 drops of mineral oil, spin briefly.

3) Run the following PCR file:

   1x3 min, 93°C;
   30x(1 min, 93°C; 1 min, 58°C; 4 min, 72°C)
   1x7 min 72°C; ->4°C.

We use a PCR buffer which gives good results for many different RT-PCRs:

Composition of the 10x buffer: 166mM $(NH_4)_2SO_4$, 670mM Tris-HCl pH 8.8, 67mM $MgCl_2$, 100mM β-mercaptoethanol, 68µM EDTA. Difficult PCRs are improved by adding 0.1U DeepVent Polymerase (New England Biolabs, Beverly, MA).

Second round (nested) PCR

For the second (nested) PCR we usually perform PCR with a less expensive polymerase: Super*Taq* (HT Biotechnology, Cambridge, England) and the supplied Super*Taq* buffer. Reaction volumes are 50µl and PCR cycles are similar to the ones used in the first PCR.

PCR of genomic DNA

Up to present, PTT analysis of two genes has been described: *BRCA1* and *APC*. The majority of the mutations in these genes lead to truncated proteins and both genes include a very large exon encoding more than 60% and 75% of the protein, respectively. Such an exon can be analyzed directly after amplification of genomic DNA using the modified PTT primers.

1) Make a premix containing the following per sample:

   5µl 10x PCR buffer (described above)
   3µl dNTPs (25mM)
   5µl DMSO (100%)
   1µl BSA (10mg/ml)
   2µl forward primer (20pmol/µl), 2µl reverse primer (20pmol/µl)
   0.4µl Ampli*Taq* (5U/µl),
   29.6µl H_2O
   2µl DNA (100ng/µl)

2) Add two drops of mineral oil, spin briefly.

3) Use the following PCR file:

   1x 3 min, 93°C;
   32x (1 min, 93°C; 1 min, 59°C; 3 min 72°C)
   1 X 4 min 72°C; ->4°C.

Analyze the obtained (RT-)PCR products on an agarose gel to estimate the amount of PCR product and to check for products of aberrant sizes.

In vitro transcription and translation

In vitro transcription and translation of the PCR-products is performed using the TnT T7-coupled rabbit reticulocyte lysate system (Promega) in the presence of L-[4, 5-^{3}H] leucine (Amersham, Buckinghamshire, UK).

The amount of T7-modified template DNA required for a 25μl transcription/translation reaction is about 200-500ng (if the PCR amplification was performed in a 50μl reaction volume, 8μl of this reaction should be sufficient for the transcription/translation reaction). We have good results also after scaling down the reaction volume to 12μl. Purification of the PCR-product is not required, however, purification of products may be necessary for more detailed investigation. The transcription/translation reaction can be performed at room temperature. Translate 0.5μl Luciferase T7 Control DNA (0.5μg/μl) as a positive control for the transcription/translation reaction.

Preparation of the reaction mixture

The components of the TnT kit are kept at -70°C. The lysate should be thawed in the hand and then put on ice. The T7 RNA polymerase should be put on ice immediately.

For each sample, prepare the following reaction mixture:

    12.5μl TnT lysate
    1.0μl TnT reaction buffer
    0.5μl TnT T7 RNA polymerase
    0.5μl amino acid mixture minus leucine (1mM)
    0.5μl RNasin ribonuclease inhibitor (40U/μl)

Transcription and translation reaction

1) Add the reaction mixture (15μl) to the crude PCR product.
2) Adjust the volume to 23 μl by adding sterile H_2O (if necessary).
3) Add 2μl ^{3}H-leucine to the reaction and mix gently.
4) Incubate at 30°C for 60 min.
5) Stop the reaction by adding 25μl SDS sample buffer.
6) Boil the sample for 5 min to denature the proteins.
7) Spin down briefly.
8) If samples are not subjected to SDS-PAGE analysis immediately they can be stored at -20°C.

SDS-PAGE analysis

Using radioactive amino acids such as leucine or methionine, SDS-PAGE analysis in combination with fluorography can reveal de novo synthesized proteins. However, it is possible to label proteins also with biotin-modified lysines. These can then be detected in Western blotting by avidin-labelled enzymes, leading to the formation of colour precipitates or chemoluminescent signals (ECL, Amersham). Thereby radioactive material may be avoided in routine diagnostic laboratories. At present we do not have much experience with these alternative detection methods, therefore only radioactive detection will be described.

- Prepare the following solutions:

2x SDS sample buffer: (loading mix)	100mM Tris-HCl pH 6.8 4% SDS 0.1% (w/v) bromophenol blue 20% glycerol
(optional)	200mM DTT or 8% (v/v) β-mercaptoethanol.
Electrophoresis buffer:	25mM Tris base (3g/l) 200mM glycine (14.4g/l) 0.1% SDS (1g/l)
30% acrylamide mix:	30% acrylamide 0.8% bisacrylamide
Other stocks:	1.5M Tris-HCl pH 8.8 1M Tris-HCl pH 6.8 10% SDS 10% ammonium persulfate (APS) tetramethylethylenediamine (TEMED, Bio-Rad).

For the analysis of proteins ranging from 12-60kD we prepare a gel of 12% polyacrylamide. Furthermore, a minigel system (MiniProtean II gel system, Bio-Rad, Hercules, CA) is used to reduce the time of electrophoresis and treatment with the chemical PPO. It also requires less reagents and the analysis (PTT and SDS-PAGE) can be performed in a single day, giving the result by the next day.

Install the gel system. First the separating gel is poured. After polymerisation the stacking gel is poured, and slots are made using a comb inserted into the stacking gel. Prepare the gel mix for the separating gel. For two minigels, 10ml is sufficient. Polymerisation is started by adding both APS and TEMED.

	Acrylamide percentage			
Buffer component (volume)	7%	10%	12%	15%
Tris-HCl pH 8.8 (1.5M; ml)	2.5	2.5	2.5	2.5
Acrylamide (30%; ml)	2.3	3.3	4.0	5.0
H$_2$O (ml)	5.0	4.0	3.3	2.3
SDS (10%; μl)	100	100	100	100
APS (10%; μl)	100	100	100	100
TEMED (100%; μl)	7	4	4	4

Pour the separating gel between the glass plates. The meniscus of the solution should be far enough below the top of the plate to allow for the length of the teeth on the comb plus 1cm. Carefully overlay the gel with water to get a sharp meniscus. Let stand for at least 30 min to allow polymerisation. After polymerisation, wash the top of the separating gel several times with water. Remove as much water as possible.

Preparation of the stacking gel (3.75%)

Tris-HCl pH 6.8 (1M; ml)	0.5
Acrylamide (30%; ml)	0.5
H$_2$0 (ml)	3
SDS (10%; µl)	40
APS (10%; µl)	40
TEMED (100%; µl)	4

Mix, pour the stacking gel and insert a cleaned comb into the solution. Avoid air bubbles underneath the comb teeth. The stacking gel is ready within 10 min.

Before running the gel, remove the comb and thoroughly wash the slots with electrophoresis buffer. Place the gel in a running apparatus and pour the running buffer onto the gel.

Be aware that nonpolymerised acrylamide is neurotoxic, wear gloves. Furthermore, use clean bottles, tubes and pipettes.

Loading and running the SDS-polyacrylamide gel

1) Load a prestained SDS-PAGE molecular weight marker in the first lane (Bio-Rad).
2) Load half of each translation reaction (25µl) on the gel.
3) Load 5µl of the Luciferase control reaction.
4) Run the gel at a constant voltage of 120V.
5) Separate the samples by electrophoresis until the dye reaches the bottom of the gel.

Detection of translation products

After electrophoresis, the gel is dehydrated using dimethylsulfoxide (DMSO), and submerged in a solution containing PPO (2,5-diphenyloxazole) as a scintillating agent. The gel is subsequently rinsed in H$_2$O, dried in a vacuum slab gel drier, and autoradiographed.

1) Remove the stacking gel and place the gel in a plastic tray.
2) Submerge the gel in DMSO for 10 min.
3) Remove the DMSO.
4) Submerge the gel in fresh DMSO for 10 min.
5) Remove the DMSO.

PROTEIN TRUNCATION TEST

6) Submerge the gel in a solution of PPO in DMSO (226g/l).
7) Incubate for 10 min.
8) Remove the PPO/DMSO solution.
9) Submerge the gel in the PPO/DMSO solution for another 10 min.
10) Remove the PPO/DMSO and wash the gel with H_2O.
11) Put the gel on 2 sheets of thin filter or chromatography paper, of which the upper sheet is wet.
12) Cover the gel with saran wrap.
13) Dry the gel in a vacuum slab gel dryer for about 1 hr at 60-70°C.
14) Autoradiograph the gel at -70°C using Kodak X-OMAT R film (expose overnight).

Primers for *APC* and *BRCA1* mutation detection

Primer sequences employed for the genomic PCR-based PTT analysis of APC Exon 15[1]

Segment (exon)	Length (kb)	Codons	Primer Sequence[2]
15A-F3b	1.8	654-1264	5'-[T7-trans]-CAAATCCTAAG-AGAGAACAACTGTC-3' 5'-CACAATAAGTCTGTATT-GTTTCTT-3'
15D-J	2.1	989-1700	5'-[T7-trans]-GATGATGAAA-GTAAGTTTTGCAGTT-3' 5'-GAGCCTCATCTGTACTT-CTGC-3'
15J-Q3b	2.2	1595-2337	5'-[T7-trans]-GCCCAGACTG-CTTCAAAATTAC-3' 5'-CTTATTCCATTTCTACC-AGGGGAA-3'
15P-3'UTR	2.4	2101-2844	5'-[T7-trans]-TGGAAAGCTAT-TCAGGAAGGTG-3' 5'-CCAGAACAAAAACCCTC-TAACAAG-3'

[1] Primer sequences will also appear in "Methods in Molecular Genetics: Human Molecular Genetics" (KW Adolph ed.) Vol 8, Academic Press.

[2] In the primer sequences listed, [T7-trans] means: 5'-GGATCCTAAT-ACGACTCACTATAGGAACAGACCACCATG-3'. Note that this primer contains a *Bam*HI restriction site at the 5' end, which is optional.

Note that the RT-PCR based PTT analysis of APC exons 1-14 has been described by Powell et al. (1993).

Primer sequences employed for genomic and RT-PCR based PTT analysis of BRCA1[3]

Fragment (Exon)	Length (kb)	Nucleotide position[5]	Primer Sequence[4]
11 A	1.33	793-813	5'-[T7-trans]-CTTGTGAATTTT-CTGAGACGG-3'
		2125-2103	5'-ATGAGTTGTAGGTTTCTG-CTGTG-3'
11 B	1.46	1921-1943	5'-[T7-trans]-ACAATTCAAAAGC-ACCTAAAAAG-3'
		3383-3359	5'-AACCCCTAATCTAAGCATA-GCATTC-3'
11 C	1.12	3061-3082	5'-[T7-trans]-CACCACTTTTTC-CCATCAAGTC-3'
		4183-4161	5'-TTATTTTCTTCCAAGCCCG-TTCC-3'
2-10	2.15	36-57	5'-GTGGGGTTTCTCAGATAA-CTGG-3'
		2125-2103	5'-ATGAGTTGTAGGTTTCTG-CTGTG-3'
2-10 nested	0.88	100-123	5'-[T7-trans]-GTTCATTGGAAC-AGAAAGAAATGG-3'
		979-959	5'-TTCTCATGCTGTAATGAG-CTGG-3'
12-24	>1.6	4011-4032	5'-TCACAGTGCAGTGAATTG-GAAG-3'
		??	5'-GTAGCCAGGACAGTAGAA-GGA-3'
12-24 nested	1.55	4153-4173	5'-[T7-trans]-AAGAAAGAGGAA-CGGGCTTGG-3'
		5693-5672	5'-GATCTGGGGTATCAGGTA-GGTG-3'

[3] Primer sequences were published before (Hogervorst et al. 1995).

[4] In the *BRCA1* primer sequences listed, [T7-trans] means: 5'-GCTAATACGACTCACTATAGGAACAGACCACCATGG-3'. Note that the T7-TRANS extension differs from the extension used for PTT analysis of *APC*.

[5] Nucleotide position according to Miki et al., 1994. Accession number U14680.

References

Hogervorst, F.B.L., Cornelis, R.S., Bout, M., van Vliet, M., Oosterwijk, J.C., Olmer, R., Bakker, E., Klijn, J.G.M., Vasen, H.F.A., Meijers-Heijboer, H., Menko, F.H., Cornelisse, C.J., den Dunnen, J.T., Devilee, P., van Ommen, G.-J.B. Rapid detection of *BRCA1* mutations by the protein truncation test. *Nature Genetics* **10**: 208-212, 1995.

van der Luijt, R., Meera Khan, P., Vasen, H., van Leeuwen, C., Tops, C., Roest, P., den Dunnen, J., Fodde, R. Rapid detection of translation-terminating mutations at the adenomatous polyposis coli (*APC*) gene by direct protein truncation test. *Genomics* **20**: 1-4, 1994.

Van der Luijt, R.B. and Meera Khan, P. (1995) Protein truncation test for presymptomatic diagnosis of familial adenomatous polyposis. In "Methods in Molecular Genetics: Human Molecular Genetics" (K.W. Adolph, ed.), Vol. 8, in press. Academic Press San Diego.

Miki, Y., Swensen, J., Shattuck-Eidens, D., Futreal, P.A., Harshman, K., Tavtigian, S., Liu, Q., Cochran, C., Bennett, L.M., Ding, W. et al. A strong candidate for the breast and ovarian cancer susceptibility gene *BRCA1*. *Science* **266**: 66-71, 1994.

Powell, S.M., Petersen, G.M., Krush, A.J., Booker, S., Jen, J., Giardiello, F.M., Hamilton, S.R., Vogelstein, B., Kinzler, K.W. Molecular diagnosis of familial adenomatous polyposis. *New Engl. J. Med*. **329**: 1982-1987, 1993.

Roest, P.A.M., Roberts, R.G., Sugino, S., van Ommen, G.-JB., den Dunnen, J.T. Protein truncation test (PTT) for rapid detection of translation-terminating mutations. *Hum. Mol. Genet*. **2**: 1719-1721, 1993.

Sarkar, G., Sommer, S.S. Access to a messenger RNA sequence or its protein product is not limited by tissue or species specificity. *Science* **244**: 331-334, 1989.

MAMA
MONOALLELIC MUTATION ANALYSIS
Nickolas Papadopoulos*, Ken Kinzler, and Bert Vogelstein
*Communicating author

In dominantly inherited diseases, germline mutations occur in only one allele and they are often masked by the normal allele. MAMA is a sensitive and specific diagnostic strategy for the identification of germline mutations. The basis of this strategy is to isolate human alleles in somatic cell hybrids with rodent cells, so that independent analysis of their expression is possible. The strategy is simple and utilizes routinely available clinical samples, such as peripheral blood lymphocytes (PBL). It is divided in three steps:

1) Establishment of rodent/human hybrids.
2) Documentation of monoallelic hybrids.
3) Analysis of their expression.

The strategy is shown schematically in Fig. 1.

Requirements

- Rodent recipient cells that provide selection for specific human chromosomes. For example rodent cells that are auxotrophic or biosynthetic mutants can be used. The human chromosome with the gene of interest needs to carry another gene that complements the defect of the rodent cells. Thus growth under selective conditions will only yield viable hybrid cells that include at least one of the human chromosomes with the gene of interest.

- Two microsatellite markers, at least one of them heterozygous in the investigated individual, one marker should be located distal and one proximal to the gene of study for documentation of the presence of the region of interest in the hybrid and for distinction between the maternal and the paternal alleles.

- Antibodies that can identify the human protein in a rodent background.

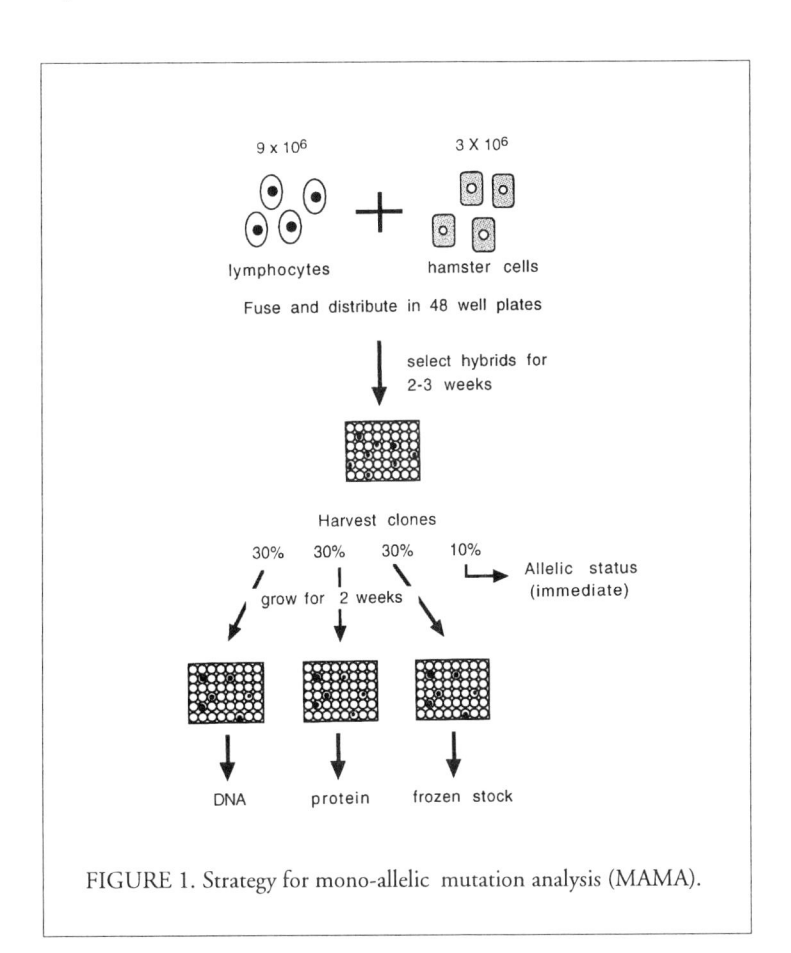

FIGURE 1. Strategy for mono-allelic mutation analysis (MAMA).

Protocol

Establishment of hybrids
Documentation of monoallelic hybrids
Protein expression analysis
Interpretation of the results

Establishment of hybrids

Preparation of human donor cells

Isolate PBL from fresh blood. PBL are isolated with Histopaque according to the protocol from Sigma. It is possible to use frozen stocks of PBL that have been stored at -80°C for up to one year. If lymphoblastoid lines are used, just spin the cells down and resuspend in growth medium.

Preparation of rodent recipient cells

Trypsinize cells and resuspend in growth medium.

Fusion of human and rodent cells

1) Count both donor and recipient cells.
2) Mix 9×10^6 human PBL with 3×10^6 rodent cells.
3) Aliquot in a 15ml conical tubes.
4) Collect cells by centrifugation at room temperature at 200x g.
5) Resuspend cell pellet in 10ml DMEM without serum.
6) Collect cells once more by centrifugation.
7) Aspirate media until the pellets are almost dry.
8) Loosen pellet by tapping the tube.
9) Slowly add 1ml of 50% PEG in PBS (Sigma) to the pellet.
10) Allow cells to fuse for 1 min at room temperature.
11) Add 10ml DMEM to terminate fusion.
12) Collect cells by centrifugation as above.
13) Resuspend the pellet in 48ml of the appropriate medium and distribute into a 48-well plate.
14) Place cells in incubator and allow them to recover.
15) Start selection two days later.
16) Replenish media every three days. This will wash away any unfused human lymphocytes.
17) Two to three weeks later visible clones are harvested by trypsinization.
18) One tenth of the cells are used for DNA isolation, the rest are split to three wells each in separate 48-well plates, one intended for DNA isolation, one for protein, and one to make a frozen stock of cells.

Documentation of monoallelic hybrids

1) Collect the one tenth of the cells from above by centrifugation in a 1.5 ml Eppendorf tube.

2) Resuspend the cell pellet in 50μl proteinase K (100μg/ml) in TE10 (10mM Tris-HCl pH 8.0, 1mM EDTA).
3) Incubate at 58°C for 1.5 hr.
4) Chill on ice.
5) Boil for 10 min.
6) Spin 5 sec.
7) Use 2-4μl for PCR.
   PCRs are carried out in 96-well plates.
   For a 10μl reaction use:
   5μl of PCR Master Mix (Boehringer Mannheim),
   2-4μl of DNA,
   1μl of end labelled primers (5ng each),
   H_2O to 10μl.
8) Analyze PCR products on a 6% denaturing polyacrylamide gel.
9) Visualize by autoradiography.
10) Score monoallelic hybrids.

Growth of monoallelic hybrids

At this point you may elect to continue propagating only a certain number of the monoallelic hybrids identified by the DNA typing. A good number is five of each allele. Continue culture for one to two weeks, or until the wells are confluent.

DNA analysis

This step ensures retention of the appropriate allele in the hybrid during the two week propagation period of the clone for cells growing in the 48-well intended for DNA isolation.

1) Wash wells with PBS or HBSS.
2) Add 100μl of proteinase K in TE10 as above.
3) Incubate the 48-well plate on a heating plate at 58°C for 1.5 hr.
4) Transfer liquid into 1.5ml Eppendorf tube.
5) Boil for 10 min.
6) Spin briefly.
7) Perform PCR as described above.

Protein expression analysis

1) To cells growing in 48-well plates intended for protein analysis, add buffer composed of 0.0625M Tris-HCl pH6.8, 5% β-mercaptoethanol, 2% SDS, 10% glycerol, 0.025% bromophenol blue.

2) Transfer to 1.5ml Eppendorf tubes.
3) Protein extracts are stored at -80°C.
4) Western blots are performed under standard conditions.

Frozen stocks

The cells growing in 48-well plates intended as frozen stocks are trypsinized, resuspended in freezing media, and stored at -80°C.

Interpretation of the results

Two results are possible:

1) Lack of full length polypeptide associated with either the maternal or the paternal allele-specific hybrids. This can be the result of many kinds of mutations, including:

- Deletions
- Insertions
- Splice site alterations
- Nonsense mutations
- Mutations that affect transcription
- Mutations that affect translation.

2) No apparent change in the length of the polypeptide in either of the allele-specific hybrids is observed. This could be the result of a point mutation, or a small deletion/insertion. Such mutations are easily identifiable in a monoallelic background by sequencing RT-PCR products.

Failure to identify a mutation would suggest that the gene analyzed may not cause the disease.

Reference

Papadopoulos, N., Leach, F., Kinzler, K., and Vogelstein, B. Monoallelic mutation analysis (MAMA) for identifying germline mutations. *Nat. Genet.* 11: 99-102, 1995.

NOVEL ASSAY FORMATS

OLIGONUCLEOTIDE ARRAYS
FOR SCANNING NUCLEIC ACID SEQUENCES

Edwin Southern*, Uwe Maskos, Stephen Case-Green, and Martin Johnson

*Communicating author

Arrays of oligonucleotides, corresponding to a full set of complements of a known sequence, can be made in a single series of base couplings in which each base in the complement is added in turn (Southern et al., 1994). Coupling is carried out on the surface of a solid support such as a glass plate, modified to allow oligonucleotide synthesis, using a device which applies reagents in a defined area. The device is displaced by a fixed movement after each coupling reaction so that consecutive couplings overlap only a portion of previous ones. The shape and size of the device, and the amount by which it is displaced at each step, determines the length of the oligonucleotides. Certain shapes create arrays of oligonucleotides from mononucleotides up to a given length in a single series of couplings. The array is used in a hybridisation reaction to a labelled target sequence, and shows the hybridisation behaviour of every oligonucleotide in the target sequence with its complement in the array. Applications include sequence comparison to test for mutations.

The series of oligonucleotides is made by coupling the bases to a solid surface in the order in which they occur in the complement of the target sequence. The reagents used to synthesise the oligonucleotide are applied in a patch to the surface of the solid substrate. (We use glass plates that have been coated with a linker derived from hexaethylene glycol (Maskos and Southern 1992). If the template used to create the patch were kept in the same place during all the synthetic cycles, the result would be a complete complement of the target sequence. If, however, the template is moved along the surface after each coupling, the result is a series of oligonucleotides, each one complementary to a region of the target sequence (Fig. 1).

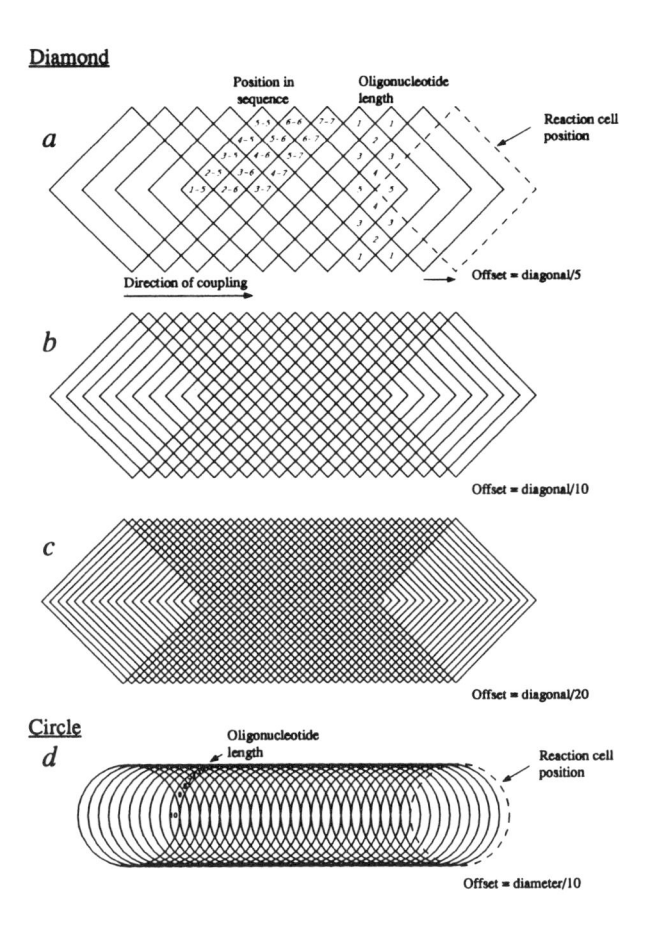

FIGURE 1. Configuration of scanning arrays. Two template shapes are illustrated here: a "diamond" shape created by turning a square through 45°, and a circle. We describe arrays made using the circular template; a diamond shaped cell can be made from aluminium or stainless steel, to be used with aminated polypropylene (Matson et al., 1993) or other flexible substrates. In either case, a cell is created by pressing a seal of the desired shape against the substrate on which synthesis takes place. Reagents and washing solutions are introduced at the bottom of the cell, and removed from the top or the bottom. These processes can be carried out automatically by coupling the cell to an automatic synthesiser such as the ABI 381A. After each coupling, the cell is moved along the substrate by a predetermined offset. For a cell of fixed dimension, longer oligonucleotides in smaller areas are made by reducing the amount of the offset. Along the centre line, the length of the oligonucleotides that are synthesised is equal to the diameter of the cell divided by the offset.

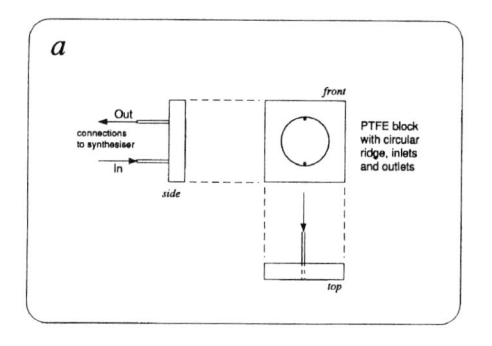

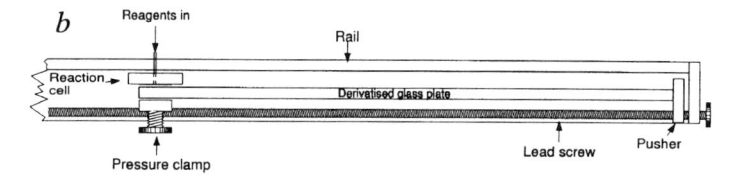

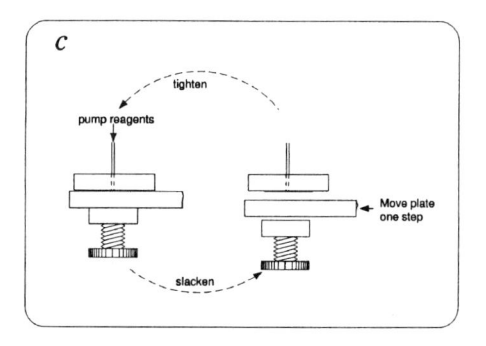

FIGURE 2. Apparatus used for applying reagents to glass plate.

a) The Teflon reaction cell consists of a block, 50x50mm, with a circular ridge, 0.5mm height 30mm diameter. Two holes, 1.0mm diameter are drilled inside the top and bottom of the circle to take the shortened 19G syringe needles used to connect the reaction cell to the oligonucleotide synthesiser. The assembled cell is fixed to the rail which carries the lead screw shown in b).

b) The apparatus is mounted on a rail, an "L" section bar, that is fixed to the front of the ABI 381A oligonucleotide synthesiser. The lead screw, 1mm pitch, is fitted with a "pusher" that drives the glass plate across the front of the reaction cell. Its knob is marked at half turns, to make it easy to drive the plate forward in half millimeter steps. The pressure clamp is a modified carpenter's "G" clamp, fixed to the back of the rail, and with a polyethylene cushion mounted on its pressure pad. Moderate hand tightening is enough to seal the glass plate against the reaction cell without breaking it.

c) One coupling cycle comprises: clamping the plate up to the reaction cell; activating the synthesiser to go through the preprogrammed cycle to couple the appropriate nucleotide; slackening the clamp and pushing the plate one offset by turning the lead screw.

Protocol

Making arrays

Hybridisation reactions

Making arrays

Glass plates (50x220-300x3mm) are first coated with a covalently attached linker (Maskos and Southern, 1992). Plates are immersed in a mixture of glycidoxypropyl trimethoxysilane/diisopropyl-ethylamine/xylene (17.8:1:69 by volume), heated to 80°C and held at this temperature for 9 hr, and then washed in ethyl acetate or acetone and ether. In a second step, the plates are heated in neat hexaethylene glycol, containing a catalytic amount of sulphuric acid, at 80°C for 10 hr, washed with ethanol and ether, air dried and stored at -20°C. Oligonucleotide synthesis is performed using standard reagents for phosphoramidite chemistry, omitting the capping step. The ABI 381A is programmed to couple bases in the order corresponding to the complement of the target sequence, with an interrupt after deprotection (Fig. 2). The scale is for 0.2μmol synthesis, adjusted slightly to provide volumes that would just fill the reaction chamber. Some steps may be shortened to speed up the procedure (Table 1).

Final deprotection in 30% ammonia is carried out in a specially constructed bomb, comprising a chamber (230x230x20mm) cut into a Nylon block (300x300x30mm), sealed by a sheet of silicone rubber (3mm thick), compressed against the rim of the chamber by clamping the whole assembly between two mild steel plates (6mm thick) using four bolts along each side of the square. After 5-8 hr at 55°C the bomb is cooled to 4°C before opening. The plate is then washed in ethanol followed by Tris-HCl, EDTA (0.01M, pH 7.8, 0.1% SDS) and ethanol, and then dried in an air stream.

Hybridisation reactions

We have used a variety of target molecules in experiments with scanning arrays: synthetic oligonucleotides labelled using polynucleotide kinase with γ^{32}P-, γ^{33}P- or γ^{35}S-ATP to tag the 5' end; RNAs labelled at the 3' end using RNA ligase with 5-^{32}P cytosine-3,5-diphosphate; or transcripts of DNA fragments made from PCR amplified fragments using T7 or SP6 polymerase to incorporate α^{32}P- or α^{35}S-UTP. All of these make good hybridisation probes, although RNA produces greater problems of intrastrand hybridisation of the target sequence. Hybridisation conditions are adjusted to the properties of the sequence to be analysed; one benefit of having all lengths of oligonucleotides is that a single set of conditions will find the optimum for each position in the sequence. We usually carry out hybridisation reactions at 4-25°C, in solutions containing 3-4.5M tetramethylammoniumchloride (TMACl) or 1.0M NaCl.

Program for ABI 381A to deliver reagents for one coupling cycle.

At the interrupt, the operator moves the plate one offset, and restarts the program.

Step No.	Funct. No.	Funct. Name	Step Time (sec.)	Step No.	Funct. No.	Funct. Name	Step Time (sec.)
1	10	#18 to waste	5	26	2	Reverse flush	5
2	9	#18 to column	30	27	9	#18 to column	15
3	2	Reverse Flush	10	28	2	Reverse flush	5
4	1	Block Flush	5	29	9	#18 to column	15
5	28	Phos Prep.	3	30	2	Reverse flush	5
6	90	Tet to Column	5	31	9	#18 to column	15
5	28	Phos Prep.	3	32	2	Reverse flush	5
6	90	Tet to Column	5	33	1	Block flush	5
7	19	B+Tet to column	3	34	33	Cycle entry	1
8	90	Tet to column	3	35	10	#18 to waste	5
9	19	B+Tet to column	3	36	9	#18 to column	20
10	90	Tet to column	3	37	2	Reverse flush	5
11	19	B+Tet to column	3	38	1	Block flush	5
12	90	Tet to column	3	39	5	Advance FC	1
13	9	#18 to column	2	40	6	Waste port	1
14	4	Wait	30	41	82	#14 to waste	5
15	10	#18 to waste	5	42	14	#14 to column	40
16	2	Reverse flush	5	43	10	#18 to waste	5
17	1	Block flush	5	44	9	#18 to column	30
18	81	#15 to waste	5	45	1	Block flush	5
19	13	#15 to column	23	46	7	Waste bottle	1
20	10	#18 to waste	5	47	2	Reverse flush	5
21	4	Wait	30	48	10	#18 to waste	5
22	2	Reverse flush	10	49	9	#18 to column	30
23	1	Block flush	5	50	2	Reverse flush	5
24	10	#18 to waste	5	52	1	Block flush	5
25	9	#18 to column	15	53	34	Cycle end	1

Hybridisation is easily done by introducing a solution of labelled target into a gap between the array plate and a plate of similar size held against it. The solution is drawn into the gap by capillary action. After hybridisation, the plate is rinsed in the hybridisation solvent and exposed through Clingfilm to a storage phosphor screen (Fuji STIII) which is then scanned in a Molecular Dynamics 400A PhosphorImager. Care must be taken not to warm the plate by contact with fingers, as this can melt short duplexes.

We find some deterioration of the plates with time. This takes the form of high background in very small spots and seems to be caused by damage to the surface, exposing the glass beneath the linker. The damage may be caused by dust particles trapped against the glass surface during hybridisation or exposure. Other indications of this kind of damage can sometimes be seen as rings corresponding to the edge of the Teflon reaction chamber, which must be caused by the high point pressure used to seal it against the glass. However, the oligonucleotides appear to be very stable; we have used some arrays for more than a year without apparent loss of performance.

Cautionary note: In many analyses it is not easy to correlate the hybridisation data with the sequence of the target. This problem arises when the target is folded in such a way that some of the sequence is "hidden" from the oligonucleotides.

References

Breslauer, K. J., Frank, R., Blocker, H., and Marky, L. A. Predicting DNA duplex stability from the base sequence. *Proc. Natl. Acad. Sci. USA* **83**: 3746-3750, 1986.

Maskos, U. and Southern, E.M. Oligonucleotide hybridisations on glass supports: a novel linker for oligonucleotide synthesis and hybridisation properties of oligonucleotides synthesised in situ. *Nucl. Acids Res.* **20**: 1679-1684, 1992.

Matson, R.S., Rampal, J.B. Coassin, P.J. Biopolymer synthesis on polypropylene supports. I. Oligonucleotides. *Anal. Biochem.* **217**: 306-319, 1993.

Southern, E.M., Case-Green, S.C. Elder, J.K., Johnson, M., Mir, K.U., Wang, L. and Williams, J.C. Arrays of complementary oligonucleotides for analysing the hybridisation behaviour of nucleic acids. *Nucl. Acids Res.*, **22**: 1368-1373, 1994.

Wetmur, J. G. DNA probes: Applications of the principles of nucleic acid hybridization. *Crit. Rev. Biochem. Mol. Biol.* **26**: 227-259, 1991.

OPTICAL WAVEGUIDE
DEVICE FOR DNA HYBRIDIZATION ANALYSIS

Julian Gordon*, Joanell V. Hoijer, Wang-Ting Hsieh, Cynthia Jou, and Don I. Stimpson

*Communicating author

The optical waveguide described here is a very simple device for direct visualization of DNA hybridization and melting off. The principle is the use of a microscope slide as a support for the oligonucleotide probes and as a waveguide. Non-coherent white light is injected into the slide end-on and is transmitted through by repeated total internal reflection from the surfaces. An evanescent wave is generated in an outer layer of the slide of approximately one-half the wavelength of the light source. Hybridization is detected using colloidal selenium of approximately 0.2µm diameter as a label. When the colloidal selenium is within the evanescent wave, it is illuminated and scatters light strongly. Beyond this boundary layer the colloid actually absorbs any stray scattered light. Extremely high concentrations of label are thus advantageous and also serve to drive the forward direction of the binding reaction. The DNA to be hybridized is either biotinylated by incorporation of biotinylated nucleotides during synthesis, or the PCR products are generated with biotinylated oligonucleotides as primers. Figure 1 below illustrates these basic principles.

Together with a thermostatically controlled heating plate, this device has been shown to be capable of monitoring specific hybridization by generating melting curves within a few minutes (Stimpson et al., 1995). Data can be collected and quantitated with an off-the-shelf video camera and digital image software.

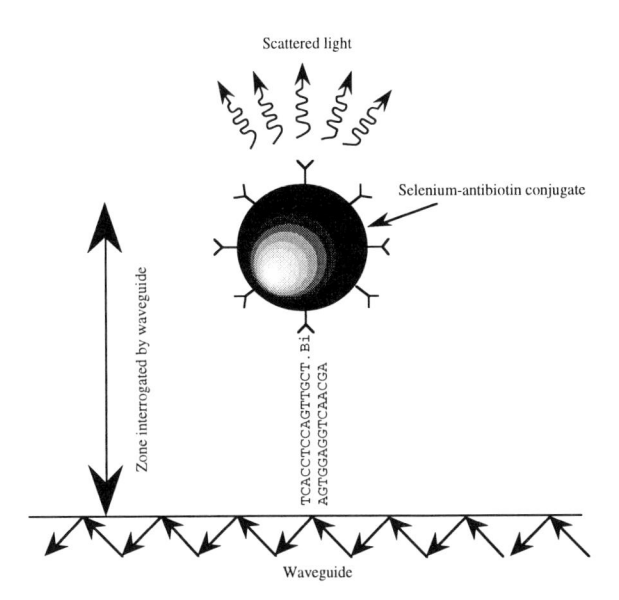

FIGURE 1. Schematic diagram of optical waveguide-illumination particle scattering-detection. Light repeatedly reflected through the body of the slide results in an evanescent wave which illuminates the exterior surfaces of the waveguide, including the sample side. An oligonucleotide attached to the surface hybridizes with its biotinylated complement. The biotinylated complement binds to the colloidal selenium, coated with anti-biotin. The colloidal selenium in the boundary layer will thus act as a scattering source and scatter light uniformly in all directions. Any selenium outside of this layer will not scatter.

Protocol

Preparation of chips

Preparation of selenium colloid

Hybridization reaction

Staining reaction

Melting curves

Preparation of chips

Glass substrates are cleaned by soaking in 2M NaOH for 1 hr, followed by rinsing with HPLC grade water (Fisher). The glass is protein coated by application of 0.05% casein in PBS (10mM phosphate, pH 7.4, 120mM NaCl, 2.7mM KCl) for 1 min. The casein solution is flushed from the surface with water from a wash bottle. The casein-coated surfaces display a "sheeting" action. All excess water is drained from the surface with careful application of the wash bottle stream.

Oligonucleotides are from commercial sources, rehydrated in 50µl water and diluted 1:20 in PBS, yielding final concentrations in the range from 1 to 15µM. Oligonucleotides are either manually pipetted on to the surface or they are applied with robotics, using drill-bits as defined pin applicators. The two sizes yield approximately 1mm or 0.5mm diameter spots, respectively. They are loosely mounted in a jig, blunt end down, and the mass of the bits defines the downward pressure on the substrate. An X-Y-Z mover (Automove 102, Asymtek, Carlsbad, CA) provides the transport in the desired arrays. The slides are allowed to dry at room temperature for at least 6 hr, and dessicated if the ambient humidity is high. Excess oligonucleotides are washed off with water.

A chamber is constructed (see Fig. 2) with a second cover slide and two strips of double-faced tape. The two slides are offset slightly, to provide a surface for the addition of samples. The waveguide slide is spray-painted with black paint on the outside surface. The aluminum heating block is placed immediately below and in contact with the slide.

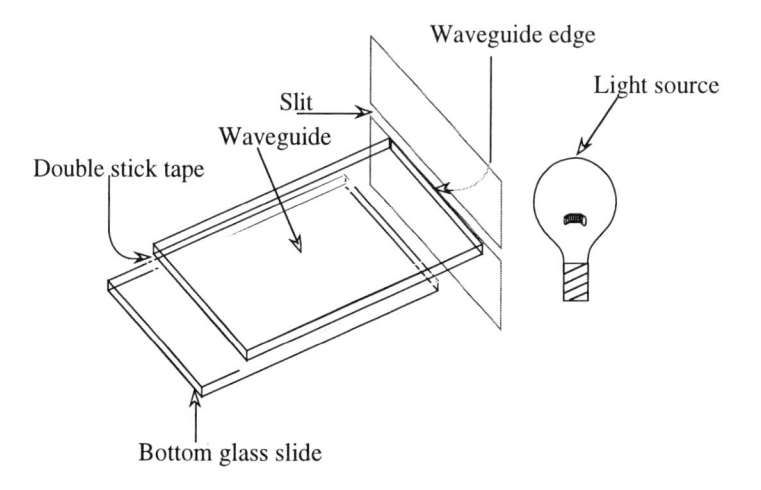

FIGURE 2. Assembly of the waveguide. The waveguide slide and the additional slide form a cuvette with the aid of two strips of double faced tape. The waveguide is illuminated end-on via the slit. The scattered light is viewed from above.

Preparation of selenium colloid

Four ml of 1% (w/v) sodium ascorbate (Sigma) is added to 200ml of boiling water. Two ml of 1% (w/v) selenium dioxide solution is added under rapid mixing (see US patent 4954452). The colloid then has a particle size of about 0.2 microns, pH 5.4, an optical density of 32, and an absorption maximum at 546nm.

Polyclonal rabbit anti-biotin is diluted to 1.13mg/ml in PBS.

A selenium-antibody conjugate is prepared fresh before use by mixing 2.5μl of the anti-biotin with 1ml of selenium colloid, followed by the addition of 30μl of a 20% (w/v) bovine serum albumin solution (Sigma).

Hybridization reaction

Solutions of 3'-biotinylated DNA are diluted into 1%(w/v) casein in TBS (10mM Tris-HCl pH 7.8, 15.4 mM NaCl). Final concentrations of DNA fragments can vary from a high of about 500nM to the lower limit of detection at 0.5nM.

The DNA solution (50μl) is applied to the offset of the lower slide and introduced into the chamber by capillary action. Incubation is typically for 5 min at room temperature (23-25°C).

Staining reaction

1) The selenium-antibiotin conjugate is diluted 1:1 with 2% (w/v) casein in TBS.
2) The casein-conjugate mixture is applied as a pool on the offset and the DNA solution drawn out of the other with a filter paper, thus displacing it.
3) The system is illuminated and video recording initiated.

Melting curves

Melting is initiated between 1 and 5 min after addition of the conjugate. Heating can be performed rapidly, and the melt is usually complete in 5 min.

Selected frames from the video recording are digitized and used to determine the intensity of scattering at each spot. The thermocouple read-out is included in the video image so that the temperature at each frame can be observed. Fig. 3 shows a typical result.

FIGURE 3. Melting curves obtained with optical waveguide device. The sequences bound to the substrate were A: 5'-AGTGGAG_G_TCAACGA-3'-NH$_2$ and B: 5'-AGTGGAG_A_TCAACGA-3'-NH$_2$. These were hybridized with the following complements or single base-mismatched (underlined) complements: A*: 5'-TCGTTGA_C_CTCCACT-3'-biotin and B*: 5'-TCGTTGA_T_CTCCACT-3'-biotin.

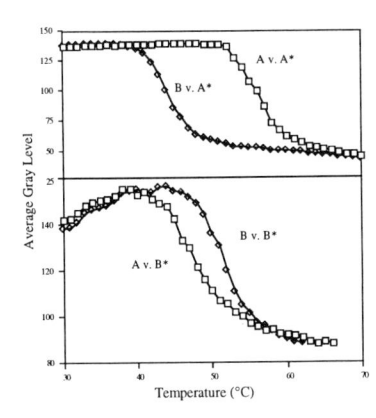

Materials

#2 microscope cover slides, Corning, Inc. or equivalent

Casein: In-house alkali-treated (see Ching et al., US patent 5,120,643, 1992)

Oligonucleotides: Purchased from Synthecell (Columbia, MD) or Genosys (Woodlands, TX). Note that under the conditions described here, amine-derivatives of the oligonucleotides (3'- or 5'-modified) give better signal than unmodified ones.

Drill-bit pins as applicators: HSS 67 or HSS 80, Hayden Twist Drill, Warren, MI

Double-faced tape: Arcare 7710B, Adhesives Research Inc., Glen Rock, PA

Selenium dioxide: Aldrich

Projection lamp: 150 watts, Dolan Jenner Fiberlite series 180

Cohu CCD camera: Model 4815, Cohu Inc., San Diego, CA

8-bit frame grabber: Imaging Technology, PC Vision Plus, Woburn, MS

NIH Image software: From FTP internet site zippy.nimh.nih.gov/pub/nih-image

Aluminum heating block: 1.5"x1.5"x0.25", containing two heating elements and a thermocouple (Watlow 965 demonstration unit, Winona, MN)

Reference

Stimpson, D.I., Hoijer, J.V., Hsieh, W.T., Jou, C., Gordon, J., Theriault, T., and Baldeschwieler, J. Real time detection of DNA hybridization and melting on oligonucleotide arrays using optical wave guides. *Proc. Natl. Acad. Sci. USA* 92: 6379-6383, 1995.

The figures were all modified from Stimpson et al., 1995.

CHAPTER · 31

CONSTRUCTION OF
MANIFOLD SUPPORTS

Jüri Parik, Arild Lagerkvist, Marek Kwiatkowski,
and Ulf Landegren*

*Communicating author

Mutation analysis frequently requires processing of large numbers of individual samples through consecutive reaction steps, involving risks of sample mix-up, contamination and tedium. A type of "molecule handles", called manifolds, can greatly simplify and streamline sample handling. Manifold supports are multipronged disposable devices, the teeth of which project into corresponding reaction wells in microtiter plates or in dedicated sequencing reaction chambers (Fig. 1). By suitably modifying the surface of these supports, large amounts of specific molecules may be bound to each of the prongs, permitting the bound molecules to be processed through successive reaction steps. The supports may also be used to finally deposit the sets of reaction products that have been obtained in a suitable analysis instrument, further reducing the need for pipetting and the risk of sample contamination.

The active surface of the supports is expanded by binding porous particles to the surface of the plastic prongs. Thereby large amount of reagents, such as avidin or streptavidin, antibodies, or oligonucleotides, can be anchored to the surface of the teeth of the manifolds. Supports modified in this manner can be used to capture biotinylated or antigenic molecules, or specific DNA strand onto the solid phases. This surface-modification technique should be applicable to a wide range of combinations of plastic supports, microparticles, and proteins or other molecules (Parik et al., 1993).

As described elsewhere in this volume, we have used supports of this type for DNA sequencing (Lagerkvist and Landegren), ligase-mediated gene detection (Samiotaki et al.), and in a single-sided PCR technique (Lagerström-Fermér and Landegren), and we have also applied the supports for immunologic applications. Here we will exemplify the technique for increasing solid-support binding-capacity by expanding the surface of plastic supports through attachment of porous particles by a protocol for coupling avidin to prongs of a manifold support.

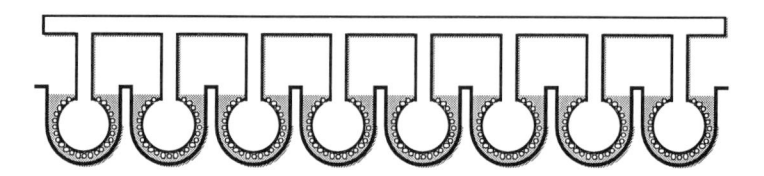

FIGURE 1. Structure of a manifold support for processing sets of molecules in e.g. mutation detection reactions and for loading the samples in an analysis instrument.

Protocol

Coupling of avidin to porous agarose particles

Annealing of agarose particles to plastic supports

Measurement of binding capacity

Coupling of avidin to porous agarose particles

1) Six ml of sedimented Sepharose particles are washed for less than one min with ice cold 1mM HCl (3x10ml) on a sintered funnel.

2) The particles are transferred to 5ml of 1.0M NaCl and 0.4M NaHCO$_3$ pH 8.3, containing 20mg of avidin.

3) The suspension is incubated rotating end-over-end for one hr, filtered, and excess active sites on the particles are blocked in 0.1M ethanolamine-HCl buffer pH 8.3 for 15 min.

4) Avidin-conjugated Sepharose particles are washed with 0.1M acetate buffer pH 4.0, followed by distilled water, and used immediately. Alternatively, batches of conjugated particles can be kept at 4°C until use in a storage buffer of 50mM Tris-HCl pH 7.3 with 0.02% (w/v) sodium azide.

Annealing of agarose particles to plastic supports

The surface of the 96-pronged support is modified by attaching avidin-coupled particles.

1) The particles are dried on the filters by washing with methanol (3x5ml), followed by triethylamine (3x5ml), and then suspended as a 50% (v/v) slurry in triethylamine.
2) The polystyrene support is washed with ethanol for 20 min in an ultrasonic bath and air dried.
3) The prongs of the support are immersed in the slurry for two seconds, air dried, and the procedure is repeated once more.
4) The supports are washed under running tap water.
5) The supports are kept in storage buffer in a humidified container at 4°C until they are used. Alternatively, they may be stored dry.

Measurement of binding capacity

It is useful to characterize the constructed supports by using a standardized assay to monitor binding capacity. We have used a competitive assay where the binding of tritiated biotin in the presence of a variable amount of unlabelled biotin is measured.

As an alternative, the binding of biotinylated oligonucleotides, also labelled with e.g. lanthanide chelates, can be determined in competition with free biotin. Bound oligonucleotides are quantitated through time-resolved fluorescence measurement in a Delfia Time-Resolved Fluorometer (Wallac OY, Turku, Finland).

Materials

Manifold supports, configured as a microtiter plate lid with 8 rows of 12 ball-and-pin extensions projecting into individual wells of a microtiter plate located underneath are available from Falcon, Oxnard CA.

The particles used for immobilization are Sepharose particles (HiTrap, NHS-activated HP Sepharose, available as a suspension in isopropanol, Pharmacia Biotech, Uppsala, Sweden).

Comments

The procedure described here is useful for obtaining a greatly expanded binding surface on plastic devices such as manifolds but also planar surfaces, microtiter wells, and dip sticks can be treated in this way. By suitable selection of organic solvents, other plastic surfaces may be modified in a similar manner. A wide selection of particles are also suitable to be used in the procedure, provided these are not dissolved in the organic solvent used. Many molecules are quite insensitive to being suspended in organic solvents. Besides avidin and streptavidin, we have immobilized protein A, antibodies, and oligonucleotides in the same way. As an alternative, molecules to be bound to the supports may be biotinylated and bound to streptavidin-coupled particles after these have been annealed to the plastic prongs.

Reference

Parik, J., Kwiatkowski, M., Lagerkvist, A., Lagerström Fermér, M., Samiotaki, M., Stewart, J., Glad, G., Mendel-Hartvig, M., and Landegren, U. A manifold support for molecular genetic reactions. *Anal. Biochem.* 211: 144-150, 1993.

MISCELLANEOUS

RED

REPEAT EXPANSION DETECTION

Martin Schalling*, Catherine Erickson-Burgess, Cecilia Zander, Kerstin Lindblad, Jeanette Johansen, and Tom Hudson

*Communicating author

The repeat expansion detection (RED) method was developed to facilitate the identification and study of dynamic mutations directly from genomic DNA. The method can be applied to detect expansion of any simple sequence repeat with a size of about 100 nucleotides and upwards. It is particularly useful in the study of expanded trinucleotide repeat sequences, such as those found in several neuropsychiatric disorders, as well as in the study of, for example, telomeric repeat sequences.

The method, originally described by Schalling et al. (1993), has been used to identify repeat expansions in genomic DNA from myotonic dystrophy patients (Schalling et al., 1993) and spinocerebellar ataxia families (Lindblad et al., in press). Several new expansions in the genome have also been described using this method (Schalling et al.; 1993, Lindblad et al., 1994) and, more recently, the method was used to associate elongated repeats with bipolar affective disorder (Lindblad et al., 1995; O'Donovan et al., 1995) as well as schizophrenia (Morris et al., 1995; O'Donovan et al., 1995).

The RED method may be of particular value in the study of disorders that display anticipation, observed as earlier age at onset or a more severe disease phenotype in later generations of a family. Anticipation is typical of all dynamic mutations isolated to date.

Contents

Basic principle of RED

The ligation reaction in RED

Gel separation, hybridization, and autoradiographic detection

Automated fragment analysis

Interpretation of RED analysis

Basic principle of RED

The RED method is based on the ability of a single oligonucleotide, composed of several tandem copies of a DNA sequence repeat motif, to hybridize to a complementary strand of the repeat sequence in genomic DNA. In the event that multiple repeat oligonucleotides hybridize at positions immediately adjacent to each other on a long, simple sequence repeat, a thermostable ligase can be used to ligate the oligonucleotides into larger single-stranded fragments. The basic principle of the strategy is illustrated in Fig. 1. By repeated cycles of denaturation and annealing of oligonucleotides, followed by ligation, fragments representing ligated copies of the oligonucleotide will be formed. The accumulation of ligation products is a linear process, with no recruitment of templates from previous rounds of ligation. The number of ligation products observed is strongly dependent upon the amount of substrate DNA used, as well as the number of cycles performed. If a high molar excess of oligonucleotide is used in the reaction, no further ligation of previously ligated products will occur, and as a consequence the maximum fragment size generated will correspond closely to the actual size of the longest repeat with a particular sequence motif in the genome.

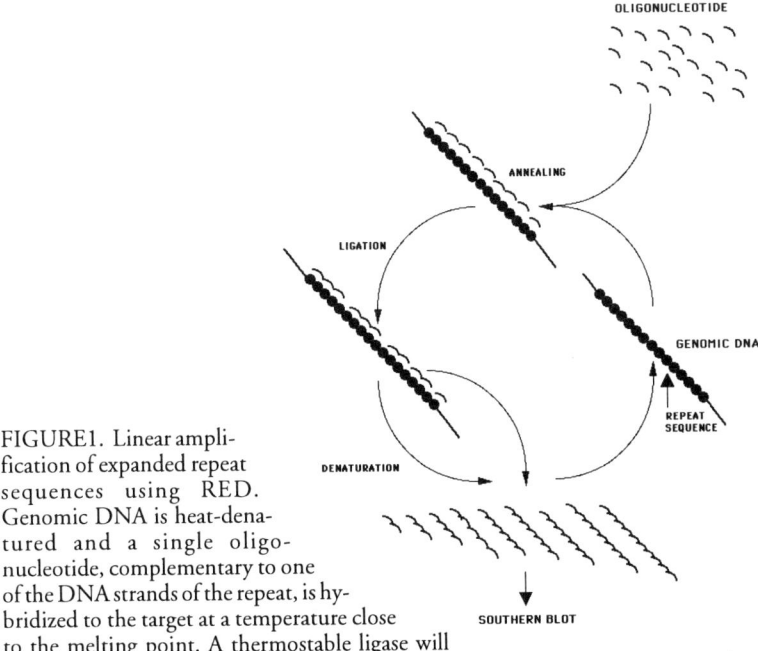

FIGURE1. Linear amplification of expanded repeat sequences using RED. Genomic DNA is heat-denatured and a single oligonucleotide, complementary to one of the DNA strands of the repeat, is hybridized to the target at a temperature close to the melting point. A thermostable ligase will covalently join adjacent oligonucleotides, producing a mixed population of single stranded DNA molecules that can subsequently be analyzed using a Southern blot procedure or by non-isotopic fragment analysis on a DNA sequencer. Detection is facilitated by using a large amount of target DNA, as well as a large number of ligation cycles. Products vary in length as multimers of the oligonucleotides used and depend upon the size of the genomic repeat sequence.

The ligation reaction in RED

Use 1-5µg of good quality genomic DNA in a total reaction volume of 10µl and a reaction mixture containing at least 50ng of phosphorylated oligonucleotide, 5U of ampligase (Epicenter Technologies), and 1µl of the supplied 10x ligation buffer, to a final volume of 10µl.

Note that the oligonucleotide must be the size of a given number of complete repeat units. For example, if the aim is to study a trinucleotide repeat expansion, a suitable probe size may be between 21-51nt but the number of nucleotides should be divisible by three, to allow copies of oligonucleotides to hybridize immediately next to each other. It is important to use pure, full-length oligonucleotides.

The cyclical ligation reactions can be performed in any thermal cycler, however, we have found that oil-free operation, such as that of the Perkin Elmer 9600, is advantageous. For optimal results, tubes with tightly fitting lids are recommended. For the Perkin Elmer instrument we recommend initial denaturation at 94°C for 5 min, followed by 500 cycles of a two-temperature cycling protocol with 10 sec at 94°C and 30 sec at a temperature very close to the melting temperature of the oligonucleotide used in the experiment. For a 30-mer $(CTG)_{10}$ oligonucleotide we have successfully used 75 or 80°C. Lower temperatures will allow for annealing and ligation also in the presence of small imperfections in the repeat sequence, which may cause an increased background. To date, the most rapid and strong signals have been obtained using a capillary thermal cycler (Idaho Technology). Using this instrument, we have been able to generate easily detectably RED products with annealing times as short as 5 sec and denaturation time set at 0 sec at 94°C, giving a total cycle time of 11 sec. One hundred such cycles, performed in about 20 min, give adequate amounts of ligation products for easy visualization following an overnight autoradiographic exposure.

Gel separation, hybridization, and autoradiographic detection

Good resolution of probe ligation products is obtained using 6-8% denaturing polyacrylamide gels where the total reaction volume is loaded using quite wide tooth combs. Following standard electrophoresis at 50°C, the gel is transferred onto an Amersham Highbond N+ membrane by simply sandwiching the wet membrane directly onto the gel with 3-4 pieces of 3MM paper on the other side of the membrane, followed by a glass plate and a weight to maximize capillary transfer of the DNA fragments. Alternatively, DNA samples can be electroblotted onto the nylon membrane in TBE for 45 min at 2.0Amp. The membrane is subsequently UV-crosslinked and can then be directly hybridised with a complementary oligonucleotide probe.

FIGURE 2: RED products identified by a Southern blot procedure. A three-generation pedigree segregating a single expanded CAG-repeat allele is shown. Individuals carrying the repeat are indicated by filled symbols. The arrow indicates 102bp, the size of a single ligation of two 51-mer oligonucleotides. The maximal sizes of RED products vary between 306-357nt, possibly indicating genomic instability and further expansion of the repeat in two of the offspring. Individuals that lack the repeat display only the 102nt product. The presence of the 102nt product in all samples indicates that successful ligation occurred in each case.

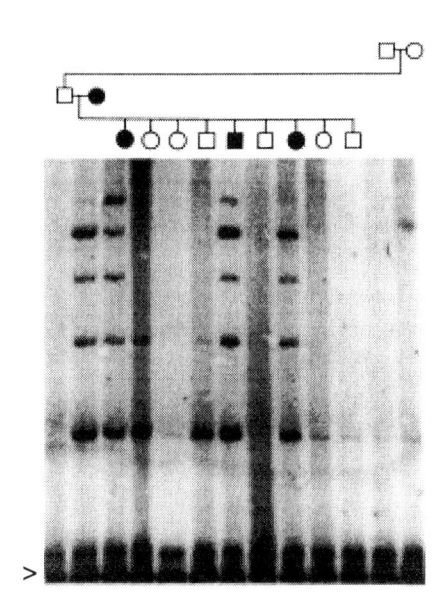

This probe is preferably labelled at the 3' end using ^{32}P α-dATP and terminal deoxynucleotidyl transferase. This labelling strategy can be used to add 20-40 ^{32}P α-dATP residues to each oligonucleotide probe, resulting in very high specific activity and short exposure times. To speed up the protocol, hybridization can be performed using Amersham's Rapid Hybe solution for 30-45 min, followed by 2 x 15 min washes as 60°C (less if an AT-rich sequence is being studied). The subsequent autoradiographic exposure time can be as short as 4-5 hr, effectively permitting complete analyses to be performed in one single day. Alternatively, conventional hybridization mixtures can be used for an overnight hybridization. An example of an analysis of the segregation pattern of a CAG-repeat is shown in Fig. 2.

Automated fragment analysis

To save time and avoid radioisotopes, we have modified the primary oligonucleotides used in the RED reaction with fluorescein to permit direct visualisation in a polyacrylamide gel, using the Pharmacia A.L.F. DNA sequencer. This mode of analysis has the advantage that results are obtained on-line from the gel, and integrated software can be used to objectively analyse any sample for the presence of long repeat segments. It has clear implications for the development of future diagnostic procedures.

Interpretation of RED analysis

- It is useful to run the gel so that also short ligation products can be analyzed. This way a fragment in the range of 80-120nt serves as a positive control, since all genomic DNA samples contain shorter repeat sequences that allow for ligation of di- or tri-mers of the oligonucleotide used. In the absence of such a "base line" ligation product, error checking should be performed. One common problem is excess salt in the DNA sample. This can sometimes be remedied by reducing the ligase buffer concentration. Low pH may also lead to reaction failure.

- Include a sample with a known repeat expansion as a positive control in any set of analyses. A reaction with no genomic DNA may also be useful, since some repeat oligonucleotides have significant self complementarity and may form concatamers at low annealing temperatures.

- The most frequently found trinucleotide repeat in the human genome is CAG/CTG. This is also the sequence motif most frequently associated with disease. It should be noted that up to 25% of normal individuals carry a single expanded allele in the range of 150-300bp in their genome. Nevertheless it is often possible to distinguish different repeats in a sample by analysing the sizes and relative amounts of ligation products in the gel. Although repeat sequences tend to increase in size from one generation to the next in a given family, it is usually easy to identify different alleles segregating in the same family.

- In association studies it is important to use ethnically matched controls, as different ethnic groups have variable levels of CAG/CTG expansion (Sirugo, personal communication).

References

Lindblad, K., Nylander, P.-O., De Bruyn, A., Sourey, D., Zander, C., Engström, C., Holmgren, G., Hudson, T., Chotai, J., Mendlewicz, J., Van Broeckhoven, C., Schalling, M., and Adolfsson, R. Detection of expanded CAG repeats in bipolar affective disorder using the repeat expansion detection (RED) method. *Neurobiol. Dis.* **2**: 55-62, 1995.

Lindblad, K., Zander C., Schalling, M., and Hudson T. M. Growing triplet repeats. *Nat. Genet.* **7**: 135, 1994.

Lindblad, K., Lunkes A., Maciel, P., Stevanin, G., Zander, C., Klockgether, T., Ratzlaff, T., Brice, A., Rouleau, G. A., Hudson, T., Auburger, G., and Schalling, M. Mutation detection in Machado-Joseph disease using repeat expansion detection. *Mol. Med.*, in press.

Morris, A.G., Gaitonde, E., McKenna, P.J., Mollon, J.D., and Hunt, D.M. CAG repeat expansions and schizophrenia: association with disease in females and with early age-at-onset. *Hum. Mol. Genet.*, **4**: 1957-1961, 1995.

O'Donovan, M.C., Guy, C., Craddock, N., Murphy, K.C., Cardno, A.G., Jones, L.A., Owen, M.J. and McGuffin, P. Expanded CAG repeats in schizophrenia and bipolar disorder. *Nat. Genet.* **10**: 380-381, 1995.

Schalling, M., Hudson, T., Buetow, K. and Housman, D. Direct detection of novel expanded trinucleotide repeats in the human genome. *Nat. Genet.* **4**: 135-139, 1993.

QUANTITATIVE PCR
COMPETITIVE PCR FOLLOWED BY QPCR DETECTION

Lucia Cavelier* and Ulf Gyllensten
*Communicating author

Competitive PCR is based on the coamplification of a target template with a dilution series of known amounts of a similar internal control sequence, for the purpose of quantitating the target sequence. The control sequence should have the same priming sites as the target so that the initial ratio of target to control products are preserved throughout amplification, and can be measured, after coamplification. Since the starting concentration of the control is known, the initial amount of target can be determined (Gilliland et al., 1990). The detection system after coamplification depends basically on the nature of the internal control used as competitor. We have used an internal control with a three base pair mismatch that allows its discrimination from the target by specific oligonucleotide hybridization, followed by electrochemiluminescence detection (QPCRTM 5000 system, Perkin Elmer Inc.). We have used this system to quantify a mitochondrial (mtDNA) deletion in brain samples of heteroplasmic Alzheimer and schizophrenic patients (Cavelier et al., 1995).

Protocol

Generation of an internal control

Competitive PCR reactions

Hybridization and binding to streptavidin-coated beads

Nonstringent wash and electrochemiluminescence detection

Plotting the data

Generation of an internal control

The internal control carrying the same sequence as the target to be measured in the sample, except for a three base-pair mismatch, is generated by PCR mutagenesis. The mutated fragment is cloned into the *Sma*I site of pUC19, transformed, and the plasmid prepared by standard procedures. The concentration of the internal plasmid control is estimated by measuring the optical density, and a dilution series is made in 1ml volumes, aliquoted and stored at -20°C. If you are to quantify different target templates in the same sample, we suggest constructing an internal control sequence containing one copy of each of the mutated target fragments of interest and use the same internal control plasmid preparation in all measurements.

Competitive PCR reactions

Prepare a master mix containing 50ng of genomic DNA (per 50µl reaction, for a mitochondrial target), 1.5mM MgCl$_2$, 100µM of each dNTP, 10% glycerol, 0,5% NP40, 0,5% Tween 20, 50mM KCl, 10mM Tris-HCl pH 8.3, 2.5U of Ampli*Taq* (Perkin Elmer, Inc.), and 5pmol of each primer, one of which is biotinylated. Aliquot 47.5µl of this mixture into a series of tubes containing 2.5µl from each tube of the dilution series of the internal control plasmid DNA (ten fold dilutions ranging from 10^2 to 10^9 copies/µl, for a mitochondrial target). The temperature is cycled 35 times between 94°C, 1 min; 52°C, 1 min; 72°C, 1 min.

Hybridization and binding to streptavidin-coated beads

The ratio of internal control to genomic target is determined by specific oligonucleotide hybridization, followed by electrochemiluminescent detection (ECL). Five µl of the competitive PCR products were hybridized in 60µl 1x PCR buffer (50mM KCl, 10mM Tris-HCl pH 8.3, and 1.5mM MgCl$_2$) to 1pmol of 5'TBR (tris 2,2*-bipyridine ruthenium (II) chelate)-labelled oligonucleotide, specific either for the target or the internal control sequence. The hybridization mixture is heated to 99°C for 5 min, followed by incubation for 15 min at RT. Then, 60µl of 4.5µm-diameter streptavidin-coated magnetic beads (Perkin Elmer Inc.), diluted to 0.5mg/ml in 1x PCR buffer, is added and the mixture incubated at RT for 30 min.

Stringent wash and electrochemiluminescence detection.

A stringent wash of 49°C (target) or 52°C (control) was performed for 5 min in a PCR machine, the samples were then transferred to 350µl of QPCRTM assay buffer (Perkin-Elmer Inc.), and the signal of the hybridizing oligonucleotide quantified by ECL using the QPCRTM 5000 system (Perkin Elmer, Inc.). The luminescence reaction is based on the oxidation of TPA (tripropylamine) and TBR (tris 2,2*-bipyridine ruthenium(II) chelate) in an electric field, resulting in the emission of light at 620nm, proportional to the amount of TBR. The stringent wash temperatures must be determined in advance for each of the hybridizing oligonucleotides in order to maximize the signal and avoid cross hybridization.

Plotting of the data

The natural logarithm of luminescence of control/target (ln C/T) was plotted as a function of the amount of internal control. A sigmoid curve is obtained and the original amount of target in the sample equals to the amount of internal control corresponding to ln C/T=0.

References

Gilliland, G., Perrin, S., Blanchard, K. and Bunn, H.F. Analysis of cytokine mRNA and DNA: detection and quantitation by competitive polymerase chain reaction. *Proc. Natl. Acad Sci. USA.* **87**: 2725-2729, 1990.

Cavelier, L., Jazin, E.E., Eriksson, I., Prince, J., Båve, U., Oreland, L., and Gyllensten, U. Decreased cytochrome C-oxidase activity and lack of age related accumulation of mitochondrial DNA deletions in the brains of schizophrenics. *Genomics.* **29**: 217-224, 1995.

CAPTURE PCR
AMPLIFICATION WITH SINGLE-SIDED SPECIFICITY ACROSS MUTATION BREAKPOINTS

Maria Lagerström-Fermér* and Ulf Landegren
*Communicating author

In order to identify breakpoints for deletions, insertions, and inversions of genomic DNA, using sequence information near one breakpoint, a method to amplify DNA segments with single-sided specificity may be used (Lagerström-Fermér et al., 1993). Similar techniques are needed when looking for different copies of genes that share a common motif and to complement partial cDNAs. A number of techniques have been established to amplify DNA segments using sequence information from only one end of the segment of interest, including inverse PCR (Triglia et al., 1988), and vectorette PCR (Riley et al., 1990). We present here a method, called capture PCR, where linker segments are first ligated to all ends of a restriction-digested DNA sample. After a number of DNA polymerase-mediated extension reactions from a specific, biotinylated primer, the extension products are captured on a streptavidin-coated solid support, such as a manifold or paramagnetic particles. After appropriate washes that reduce the complexity of the DNA sample, the desired fragments may be specifically amplified using another specific primer, hybridizing downstream of the biotinylated primer, in conjunction with a primer, representing the sequence added to all restriction ends by ligation (Fig. 1) (Lagerström et al., 1991). The identity of the obtained amplification products may be confirmed through another amplification, using a third, downstream primer and the linker-primer. Amplification primers can be designed to be partially overlapping in sequence and we have used the method to specifically amplify homologous genes from different genomes, using primers derived from as little as 32 contiguous, conserved nucleotide positions.

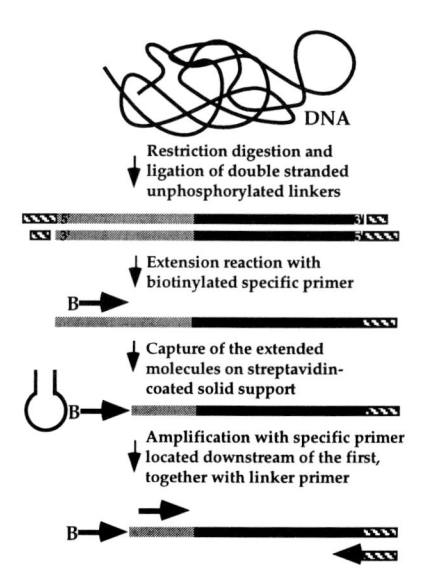

FIGURE 1. The capture PCR procedure to amplify DNA with single-sided specificity. Genomic DNA is restriction-digested and linkers are added to the ends by ligation. Next, a series of extension reactions are performed using a biotinylated primer, and extension products are trapped on a support. After washes, specific fragments are amplified by PCR, using one primer hybridizing downstream of the biotinylated one, together with another primer representing the sequence added by ligation.

Protocol

Restriction digestion and linker ligation

Biotinylated primer extension reaction

Capture of biotinylated extension products

Amplification of captured sequences

Restriction digestion and linker ligation

Genomic DNA is restriction enzyme-digested and, in the same reaction, linkers are ligated to the DNA fragments. For the combined restriction and ligation reaction 100 ng of DNA is incubated with a pair of linker oligonucleotides at 2µM each, one end of which can join to restriction ends without recreating the recognition sequence. Both oligonucleotides have 5' hydroxyl groups. The sequence of oligonucleotide pairs used to ligate to blunt ended restriction fragments is as follows:

Linker 1: 5'GCGGTGACCCGGGAGATCTGAATTC 3'

Linker 2.   3'C TAGACTTAAG 5'

1) Add the following to an Eppendorf tube:

   100ng DNA (1µl of a 100ng/µl DNA sample)
   3U restriction enzyme (e.g. *Alu*I or *Rsa*I; 0.5µl)

20pmol Linker 1 (1µl of a 20µM stock solution)
20pmol Linker 2 (1µl of a 20µM stock solution)
5mM ATP (0.5µl of a 0.1M stock solution)
4U T4 DNA ligase (0.5µl)
1µl 10x reaction buffer (100mM Tris-acetate pH 7.5, 100mM magnesium acetate, 500mM potassium acetate).
H_2O to 10µl

2) Incubate at 37°C for 3 hr or over-night

Biotinylated primer extension reaction

Multiple rounds of extension reactions are performed from a specific 5' biotinylated primer.

1) Add the components listed below to wells of a Falcon polyvinyl microtiter plate or an Eppendorf tube:

   0.5µl of the ligation mix
   2pmol biotinylated specific extension primer (1µl of a 2µM stock solution)
   2µl of 10x PCR buffer (500mM Tris-HCl pH 8.3, 10mM $MgCl_2$, 500mM KCl, 125µg/ml bovine serum albumine, dNTP 2mM each)
   0.5U *Taq* polymerase (0.1µl of a solution 5U/µl)
   H_2O to 20µl.

2) Overlay with two drops of mineral oil

3) Vary the temperature in a thermal cycler, depending on the primers used and the expected fragment sizes, for instance as follows: 94°C for 1 min, 60°C for 1 min, and 72°C for 2.5 min, repeated in 35 cycles.

Capture of biotinylated extension products

Immobilization on manifold supports

The biotin-labelled extension products are immobilized on a manifold polystyrene device, configured as a microtiter plate lid with 8 rows of 12 pin-and-ball extensions, projecting into individual microtiter wells (Falcon 3931). The prongs of the support are coated with avidin or streptavidin, as described by Parik et al. in this volume and (1993).

1) 20µl of solution A (0.1M Tris-HCl pH 7.5, 1M NaCl, 0.1% TritonX-100) is added to each extension reaction in the wells of the microtiter plate.

2) The streptavidin-coated manifold is inserted into the wells of the microtiter plate.

3) Incubate on a shaking platform at room temperature for 1 hr or over-night.

4) The solid supports are subsequently washed in individual microtiter wells once with 100µl solution A, once with a denaturing solution of 0.1M NaOH, 1M NaCl, 0.1% TritonX-100 (solution B), and again with solution A. Finally, the solid supports are washed in dH$_2$O.

Immobilization on paramagnetic particles

As an alternative, streptavidin-coated paramagnetic particles (Dynal A.S., Norway) can be used to trap the extension reaction products.

1) 40µl of streptavidin-coated paramagnetic beads (Dynal A.S.), previously washed in solution A, are added to the extension reactions.

2) Incubate 30 min on a shaking platform at room temperature.

3) Remove the supernatant in a magnetic rack, and wash the beads once with 100µl of solution A, once with solution B, and once with solution A, followed by one wash in dH$_2$O. Resuspend the particles in 10µl dH$_2$O.

Amplification of captured sequences

The immobilized extension products are used as templates in a PCR. The 50µl amplification reactions contain the oligonucleotide linker 1 and a specific primer, representing a sequence located downstream of the biotinylated primer, at 1µM each. The temperature is varied as described under 2) above.

Outer PCR

1) Add the following to a microtiter well:

   2.5µl isolated paramagnetic beads, resuspended in 10µl 1x PCR buffer, or insert one pin-and-ball extension from the streptavidin-coated manifold in each well.
   50pmol Linker 1 (2.5µl of a 20µM stock solution),
   50pmol PCR primer 1 (2.5µl of a 20µM stock solution),
   5µl 10x PCR buffer (1.5mM MgCl$_2$) ,
   1.5U *Taq* polymerase (0.35µl),
   H$_2$O to 50µl.

2) Overlay the sample with 2 drops of mineral oil.

3) Amplify in a thermal cycler, for instance as follows: 94°C for 1 min, 55°C for 1 min, and 72°C for 2.5 min, repeated in 35 cycles. When using the manifold support, remove this after the first two cycles.

4) Load 10µl of each sample on a 1.5% agarose gel. Run the gel at 150-200V for 1-2 hr. Excise the bands from the gel for later DNA sequence analysis.

Inner PCR

The identity of the amplified fragment is usually confirmed by performing an inner PCR using a third specific primer located downstream of the other two, together with a linker primer. As a control at this step, it is useful only to add the linker primer in a separate PCR. If a product appears without the addition of a specific primer, then this product is false, having linker sequence information at both ends.

1) Add the following to each microtiter well:

    1µl sample from PCR I (the sample can be taken directly from PCR I or the products of PCR I may be separated by agarose gel electrophoresis and an excised band is melted in 400µl H_2O),

    100pmol Linker 1 (5µl of a 20µM stock solution),

    100pmol PCR primer 2 (5µl of a 20µM stock solution),

    10µl 10x PCR buffer (1.5mM $MgCl_2$), 3U *Taq* polymerase (0.7µl), H_2O to 100µl.

2) Overlay the sample with mineral oil.

3) Amplify as above but for 17 cycles.

Comments

There are some critical factors to point out:

- The risk for contamination is considerable since capture PCR involves sequential amplification steps. In order to avoid contamination, tips with barriers are used for pipetting throughout the protocol. If paramagnetic particles are used, these must be washed and pipetted with extra care. To prevent contamination always open Eppendorf tubes with a fresh Kleenex - one for each tube.

- The inner PCR greatly increases the specificity of the reaction so it is recommended that, if possible, three specific primers (including one biotinylated) are used in capture PCR, located head to tail but possibly overlapping in sequence. The third primer also serves to confirm the identity of amplification products that arise in the outer PCR, since their size relationship with products generated in the inner PCR can be predicted from the relative location of the two primers in the target sequence.

- The inner PCR can also be used to incorporate a biotinylated version of the linker oligonucleotide that is added by ligation to restriction ends and used in PCR. Using this biotin, amplification products may be immobilized on a support for solid-phase sequencing. Furthermore, the specific primer used in the inner PCR can be designed to include a 5' sequence extension, for instance representing a M13 sequencing primer. In this manner amplification products may subsequently be analyzed by solid-phase fluorescent sequencing as described elsewhere in this volume by Lagerkvist and Landegren (see also Lagerkvist et al., 1994).

References

Lagerkvist, A., Stewart, J., Lagerström, M., Parik, J., and Landegren, U. Manifold sequencing: rapid processing of large sets of sequencing reactions. *Proc. Natl. Acad. Sci. USA* **91**: 2245-2249, 1994.

Lagerström, M., Parik, J., Malmgren, H., Stewart, J., Pettersson, U., and Landegren, U. Capture PCR: efficient amplification of DNA fragments adjacent to a known sequence in human and YAC DNA. *PCR Meth. Applic.* **1**: 111-119, 1991.

Lagerström-Fermér M, Pettersson U, and Landegren U. Molecular basis and consequences of a mutation in the amelogenin gene, analyzed by Capture PCR. *Genomics* **17**: 89-92, 1993.

Parik, J., Kwiatkowski, M., Lagerkvist, A., Samiotaki, M., Lagerström, M., Stewart, J., Glad, G., Mendel-Hartvig, M., and Landegren, U. A manifold support for molecular genetic reactions. *Anal. Biochem.* **211**: 144-150, 1993.

Riley, J., Butler, R., Ogiline, D., Finniear, R., Jenner, D., Powell, S., Anand, R., Smith, J.C. and Markham, A. Novel, rapid method for the isolation of terminal sequences from yeast artificial chromosome (YAC) clones. *Nucl. Acids Res.* **18**: 2887-2890, 1990.

Triglia, T., Peterson, M.G., and Kemp, D.J. A procedure for in vitro amplification of DNA segments that lie outside the boundaries of known sequences. *Nucl. Acids Res.* **16**: 8186, 1988.

INDEX

INDEX

Numbers in parentheses are the chapter numbers where the entries occur

Allele separation	(4,5,28)	Chromatin fiber	(25,26)
Allele-specific amplification	(16)	Circularizable probe	(26)
Allele-specific hybridization	(15,29,30)	Colloidal selenium	(30)
Allele-specific oligonucleotide	(15,29,30)	Competitive PCR	(33)
Allele-specific PCR	(16)	Constant denaturant capillary electrophoresis	(8)
Amplification refractory mutation system	(16)	Cot-1 DNA	(24,25)
Anticipation	(1,32)	Databases	(1-3)
AP endonuclease	(14)	dbEST	(3)
Apurinic/apyrimidinic endonuclease	(14)	Delayed-type fluorescence	(19)
ARMS	(16)	DELFIA	(19)
Arrays	(29-31)	Denaturing gradient gel electrophoresis	(7)
ASA	(16)	DGGE	(7)
ASO	(15,29,30)	DNA array	(29,30)
ASPCR	(16)	DNA chip	(29,30)
Assay formats	(29-31)	DNA fiber	(25)
Autoload	(23)	DNA methylation	(1)
Automation	(3,22)	DNA repair	(14)
Capillary electrophoresis	(8)	DNA sequencing	(22,23)
Capture PCR	(34)	Dot blot	(15)
CCM	(10-12)	Double PASA	(16)
CDCE	(8)	Dual-color	(19)
Chemical cleavage of mismatch	(10-12)	Dynamic mutations	(32)

INDEX

Ectopic transcription	(12,27)	Human genome workshop	(3)
EMC	(13)	Hybrid cells	(28)
Enzyme mismatch cleavage	(13)	Illegitimate transcription	(12,27)
EST	(3)	In situ hybridization	(24-26)
Europium	(19)	Instabile mutations	(32)
Evanescent wave	(30)	Intrastrand hybridization	(4-6)
Expressed sequence tags	(3)	Inverse PCR	(34)
FAMA	(12)	Lanthanide chelates	(19)
FISH	(24-26)	LCR	(20)
Fluorescence-assisted mismatch analysis	(11)	Ligase chain reaction	(20)
Fluorescent in situ hybridization	(24-26)	Ligase-mediated gene detection	(19,20)
Forward dot blot	(15)	Liquid-phase hybridization	(15)
Founder effect	(1)	Loss-of-function mutation	(1)
Functional assay	(1)	Low-stringency single specific primer PCR	(9)
Gain-of-function mutation	(1)	LSSP-PCR	(9)
GAWTS	(6)	MAMA	(28)
GC-clamp	(7)	Manifold	(19,23,31)
Gene signatures	(9)	Melting temperature	(7,15,29,30)
Genome Digest	(3)	Meltmap	(7,8)
Genome project	(3)	Metaphase	(24,26)
Genomic amplification with transcript sequencing	(6)	Methylation	(1)
Genomic imprinting	(3)	Microsatellite	(28)
Genotype	(15)	Minisequencing	(17,18)
Germ-line mutation	(28)	Mismatch analysis	(7,8,10-14)
Guthrie spot	(20,21)	Mismatch repair enzyme cleavage	(14)
Haplotype	(16)	Missense mutation	(1)
Heteroduplex	(5,10-14,21)	Monoallelic mutation analysis	(28)
Heteroplasmic	(17,33)	MREC	(14)
Heterozygote advantage	(1)	Multiplex	(12,18)
Hexaethylene glycol	(26,29)	Mutation databases	(2)
HGM	(3)	Mutation enrichment	(8)
HGW	(3)	MutY	(14)
HUGO	(1-3)	Nonsense mutation	(28)
Human genome mapping	(3)	OLA	(19)
Human diversity project	(1,3)	Oligonucleotide array	(29,30)
Human genome organization	(3)	Oligonucleotide chip	(29,30)
Human genome project	(3)	Oligonucleotide ligation assay	(19,20)
		Optical waveguide	(30)

191

Padlock probe	(26)
PASA	(16)
Patents	(3)
PCR amplification of specific alleles	(16)
Positional cloning	(3)
Primer extension	(17,18)
Probe arrays	(29,30)
Promoter	(27)
Protein truncation test	(27)
PTT	(27)
QPCR	(33)
Quantitative PCR	(33)
Rare sequence variant	(8,16,17,19)
Reading frame	(27)
RED	(32)
Repeat expansion detection	(32)
Resolvase	(13)
Reverse dot blot	(15)
Reverse transcription	(6,11,27)
Robotics	(22)
Scanning for mutations	(4-8,10-14)
Screening for mutations	(15-21)
SCW	(3)
Sequence analysis	(22,23)
Sequence patterns	(9)
Sequence-specific oligonucleotide	(15,29,30)
Sequencing	(22,23)
Single chromosome workshop	(3)
Single-sided PCR	(34)
Single strand conformation polymorphism	(4-6)
Somatic cell hybrids	(28)
Somatic mutation	(1)
SSCP	(4-6)
SSO	(15,29,30)
Stop codon	(6)
T4 endonuclease VII	(13)
Terbium	(19)
Tertiary structure	(4-6)
Tetramethylammonium chloride	(5,29)
Thymine glycosylase	(14)
Time-resolved fluorescence	(19)
T_m	(15)
TMACl	(5,29)
Topoisomerase I	(14)
Transition	(14)
Transcription	(27)
Transversion	(14)
Translation	(27)
TRF	(19)
Trinucleotide repeat	(1,32)
Tyramide	(24)
UHG	(30)
Universal heteroduplex generator	(30)
Vectorette	(34)
Wallace rule	(11,15,16)